James Usher

Genealogical and Historical Record of the Carpenter Family

James Usher

Genealogical and Historical Record of the Carpenter Family

ISBN/EAN: 9783337366858

Printed in Europe, USA, Canada, Australia, Japan

Cover: Foto ©berggeist007 / pixelio.de

More available books at **www.hansebooks.com**

GENEALOGICAL AND HISTORICAL

RECORD

—OF THE—

CARPENTER FAMILY,

—WITH—

A Brief Genealogy of some of the descendants of William
Carpenter, of Weymouth, and Rehoboth, Mass.,
William Carpenter, of Providence, R. I.,
Samuel Carpenter, of Penn.,
and
Ephraim, Timothy and Josias Carpenter, of Long Island.

INCLUDING

A FULL, COMPLETE AND RELIABLE
HISTORY OF THE CARPENTER ESTATE OF ENGLAND.

— BY —

JAMES USHER,
9 Murray Street, New York City.

CARPENTER FAMILY OF PENNSYLVANIA.

The persecution of Friends in England commenced about the year 1648, and reached its height during the reign of Charles II, when the prisons were filled with victims, without regard to sex, age or condition, and shiploads were banished from the Kingdom. The large accession of an industrious and thrifty population to the Island of Barbadoes, through this cause, speedily developed its natural resources, and induced others voluntarily to repair thither. Among them, it is believed, was Samuel Carpenter.

The time of his arrival can only be conjectured. He was born in 1650, fifteen years prior to the general banishment of 1664–5. According to Besse, in 1673, he suffered (in Barbadoes) considerably in distraints for refusing to bear arms.

He had then reached his twenty-third year; and it is quite probable that this difficulty with the authorities occurred soon after his arrival. The opinion that he voluntarily repaired to Barbadoes is fortified by the circumstance of his possessing ample wealth; for had he been proscribed (as in the case of Charles Lloyd and others) his property, most likely, would have been placed under præmunire.*

It is a matter of history that many Friends accumulated great wealth, with which came influence and social position. They became, not unfrequently, the associates and rivals of nobles and statesmen; they found themselves in great assemblies, sitting at the side of dignitaries of the Church, who had seats in the House of Lords, and participated in national legislation.

Their property was in real estate, or such personal effects as attracted the eye of the tax-gatherers, and easily subjected to distraint. The Friends by their principles were bound to resist the payment of tithes and the performance of military duty, and did so to the damage of their worldly estates, and too often personal liberty. Samuel Carpenter joined William Penn, in Philadelphia, 1682, where, in addition to many responsible official duties, he engaged largely in foreign commerce. He died at his original mansion,† in King (now Water)

---

* Introducing or acknowledging a higher power in the land, and creating *imperium in imperio*, by paying that obedience to Papal authority which belonged to the King. It was charged that Friends acknowledged allegiance to spiritual convictions, rather than kingly authority. The penalty placed the offender out of the King's protection, his possessions were forfeited to the King, and his body remained in prison at the King's pleasure, or during life.

† This house was subsequently occupied by his son Samuel.

Street, April 10, 1714, in the 64th year of his age. Samuel Carpenter is referred to in terms of regard by William Penn, in a letter addressed, in 1684, to Thomas Lloyd, President of Council of State. He was one of the Trustees of Public Schools, established by Friends in Philadelphia, in 1689, and also a Member of the Provincial Assembly. In 1701 Penn appointed him Member of the Council of State. He appears to have been constantly employed with public affairs, either as member of the General Assembly, Council of State, or Treasurer of the Province. The following notice of his death is taken from " Proud's History of Pennsylvania."

"In the year 1714, died Samuel Carpenter, the Treasurer of the Province, who was succeeded in office by Samuel Preston. Samuel Carpenter arrived early in the Province, and was one of the most considerable traders and settlers of Pennsylvania, where he held for many years some of the greatest offices in the government, and throughout great variety of business preserved the love and esteem of a large and extensive acquaintance. His great abilities, activity and benevolent disposition of mind in divers capacities, but more particularly among the ' Friends,' are said to have rendered him a very useful and valuable member, not only of that religious society, but also of the community in general."

There is no way of ascertaining the extent of his possessions, but the following items are incidentally alluded to in Watson's Annals.

1. A large property now covered by the town of Bristol, Pa., with extensive saw and grist mills.

2. The " Slate Roof House " on Second Street, Philadelphia. Governor Penn resided in this house in 1700, and it was afterwards owned by William Trent, the founder of Trenton. John, the eldest son of William Penn, was born here. In 1696 the Assembly of the Province met in this house. It was subsequently occupied by the officers of the 42d Highlanders, and also by those of the Royal Irish. Baron de Kalb, who fell at the battle of Camden, S. C., during the Revolution, was an inmate. Governor Forbes, the associate of General Braddock, died here. In 1868 the old mansion was demolished, and its site occupied by the Commercial Exchange.

3. Certain lots on the north side of Market Street, Philadelphia, and reaching half way to Arch Street, bounded at extremities by the Delaware River and Wood Street.

4. He was joint proprietor, with William Penn, of a grist mill on the site of Chester—the third mill in the province.

5. A lot extending from the river to Second Street, and from Norris Alley to Walnut.

6. A crane, bakery, and mansion house on the wharf. Also a store-house and grocery, and a tavern called the " Globe."

7. Half of a mill at Darby, and a saw-mill, with a pond covering 300 acres.

8. Five thousand acres in Poquassing Creek, fifteen miles from Philadelphia.

9. The island in Delaware River, opposite Bristol—350 acres.

10. An estate of 380 acres, called "Sepviser Plantation," a part of Fairhill, at the north end of Philadelphia.

11. One thousand acres of land in Pilesgrove, Salem County, N. J., part of which he sold in 1700 to John Wood.

12. Fifty acres in New Jersey, opposite Philadelphia.

13. Six hundred acres in New Jersey, on the river, bounded in part by south branch of Timber Creek.

14. Eleven hundred acres in Elsinborough. Salem County, N. J., situated near the Swede's Fort. The farm now owned by Clement Hall is part of this tract. The original purchase was made in 1696.

15. Three-sixteenths of five thousand acres of land, and a mine, called Pickering's Mine.

16. A coffee house (at or near Walnut and Front Streets, Philadelphia,) and scales.

He was actively engaged in foreign commerce, and owner in full or part of numerous vessels trading to the West Indies, and various parts of the world.

On the 12th October, 1684, Samuel Carpenter, Sen., married Hannah Hardiman, a native of Haverford, West South Wales, Great Britain. She was born in 1646, and having joined the Society of Friends, emigrated to Pennsylvania, where she became a minister of that persuasion. She died May 24, 1728, aged 82 years. A memoir of her character and services, published in Bevan's Collection of Memoirs, speaks of her as a most exemplary woman.

"Hannah Carpenter was born in Haverford-West, South Wales, where she was convinced of the principles of Friends, and where, it is said, she became very serviceable to those who were in bonds for Christ's sake. After her settlement in Pennsylvania, she was united in marriage to Samuel Carpenter, of Philadelphia, a Friend of considerable influence in the Province. Her Gospel ministry was attended with much Divine sweetness, and was truly acceptable and edifying.

She was a tender, nursing mother in the Church, and a bright example of Christian meekness. Her decease took place in 1728, at the advanced age of eighty-two years."—BOWDEN'S HISTORY OF FRIENDS IN AMERICA.

The following extract from an article in Philadelphia Commercial List, published a few years since, speaks more particularly of Samuel Carpenter as a merchant:

SAMUEL CARPENTER.

The curious view of Philadelphia, by Peter Cooper, which hangs in the Philadelphia Library, and is supposed to have been painted about the year 1714, contains, as a conspicuous object, the storehouse of Samuel Carpenter, situate upon the wharf, below Chestnut Street.

" Carpenter's Stairs," nearly opposite, was a passage from Front Street to what was then called King Street, but which, since the Revolutionary war, has been called Water Street. Carpenter's Wharf was a well-known landmark among the drab-coated men who came over with Penn, and Samuel Carpenter has literally the distinction of having been one of our first merchants. It is impossible at this time to give much information in relation to the state of our commerce during the period between the settlement of the city, in 1682, and the death of Samuel Carpenter, in 1714; but all accounts agree that Carpenter was the most successful merchant of his time. Commerce was then most confined to coasting trade, with greater voyages occasionally to the English West India Islands. Barbadoes and Jamaica were the principal points of intercourse, and from these islands came many of the settlers, whose blood still courses through our Philadelphia families.

Our exports were mostly agricultural products, in which grain, flour and tobacco held a large proportion.

Skins and furs were important articles of trade also. Ships were then more plentiful than they are now; but these ships were small craft of from one hundred to two hundred tons burthen. There was much danger from pirates, even in the short voyages which those vessels made, and the names of Kidd and Blackbeard are yet remembered

Joshua, brother of Samuel Carpenter, built Græme Hall, where, in 1856, stood the Philadelphia Arcade.

He was one of Penn's commissioners for the sale of property, and in 1708 represented the city of Philadelphia in the Provincial Assembly. He was also one of the first Aldermen appointed under the charter of 1701. His burial place was the centre of what is now known as Washington Square. Joshua was an Episcopalian. He is said to have removed to Lancaster County, Pennsylvania. Some of his descendants settled in Western Pennsylvania, and others in Kent County, Delaware.

CHILDREN OF SAMUEL AND HANNAH CARPENTER.

I.—Hannah, born 1686, married William Tishbourne, 1701, and died 1742. Her husband was mayor of Philadelphia, 1719-1720.

II.—Samuel (2d), born in Philadelphia, February 9, 1688; married Hannah, daughter of Samuel Preston (and granddaughter of Thomas Lloyd), 1711. She was born 1693 and died 1772.

III.—Joshua; died in infancy.

IV.—John; born 1690; married Ann Hoskins, 1711, and died 1724. His wife died 1719. (The descent of this branch is given elsewhere.)

V.—Rebecca, born 1692, died 1713.

VI.—Abraham, died 1702.

DESCENT OF SAMUEL CARPENTER, JUNIOR.

Samuel Carpenter, Jr., was a merchant of Philadelphia, and employed in the affairs of Provincial Government. He married Hannah Preston, 1711, and left five children—Samuel, Rachel, Preston, Hannah and Thomas.

I.—Samuel (3d), died in Jamaica, 1747, leaving three children—Samuel, Hannah and Thomas. He was a merchant, residing in Kingston. His two sons were educated in Edinburgh, and died in Kingston. Thomas left nine children—four boys and five girls.

II.—Rachel, born 1716, died 1794; unmarried.

III.—Preston, born 1721; died October 20th, 1785. He married,

1742, Hannah, daughter of Samuel Smith, of Salem County, N. J. She was born 1723. He married, secondly, Hannah Mason, but left no heirs.

IV.—HANNAH, married Samuel Shoemaker, 1746, and died 1766.

V.—THOMAS, died 1770, unmarried.

DESCENT OF PRESTON AND HANNAH CARPENTER.

1, Hannah. 2, Samuel Preston. 3, Elizabeth. 4, Rachel, *died young.* 5, Mary. 6, Thomas. 7, William. 8, Margaret. 9, John, died young. 10. Martha,

HANNAH CARPENTER, daughter of Preston and Hannah Carpenter, born 1743, and died 1820, married Charles Ellet, 1768, of New Jersey.

III.—ELIZABETH CARPENTER, daughter of Preston and Hannah Carpenter, married Ezra Firth, of Salem County, N. J.

MARY CARPENTER, daughter of Preston and Hannah Carpenter, married Samuel Tonkins. She was born 1750, and died 1821.

VI.—THOMAS CARPENTER, son of Preston and Hannah Carpenter, born 1752. Married Mary Tonkins, and left Edward, who was born 1776, and married Sarah, daughter of James Stratton, M.D., of Swedesborough, sister of Ex-Governor Charles C. Stratton. Edward Carpenter died 1813, and left—

1. Thomas P. Carpenter (died 1876), Counselor at Law, and Ex-Justice New Jersey Supreme Court. Married Rebecca, daughter of Dr. Samuel Hopkins, of Philadelphia. Children: Susan, Mary, Anna S., Thomas P. and James H. Carpenter.

2. Mary T. Carpenter, married Richard W. Howell, Counselor of Supreme Court, New Jersey.

3. James S. Carpenter, married Camilla Sanderson. Heirs: John T., Sarah S., Sophia C., Cornelia M., James E. and Preston, Camilla, Mary H., Richard H.

4. Samuel T. Carpenter, married Frances Champlain, of Connecticut. Heirs: Samuel C. B. and Frances Mary, and others (five) by second wife, Emily Thompson.

5. Edward Carpenter, married Anna Maria Howe. Heirs: Lewis H., Colonel U. S. A., James E., Sarah C., Mary H., Casper W. and Thomas P. and Charles C. S., Henrietta H.

VII.—WILLIAM CARPENTER, son of Preston and Hannah Carpenter, born 1754, died 1837. Married, first, Elizabeth Wyatt (born 1764, died 1790). Their children were: 1. Mary Wyatt, born 1783, died 1836. (She married James Hunt, of Penna., and left two sons, John and William, and daughters, Mary, Naomi and Hannah.) 2. Hannah, born 1785, died young.

William Carpenter married secondly, 1801, Mary Redman, born 1779, and died 1846. Their heirs:

1. William, married first, Hannah Scull, and secondly Phebe Warren.

2. John Redman, unmarried. He was a young man of unusual talents, and at the time of his death (1833) Cashier of the Branch Bank of the United States, at Buffalo, N. Y.

3. Rachel R., married Charles Sheppard. Heirs: William C. (who married Hannah E. Tornes), and John R. C.

4. Hannah, died young.

5. Samuel Preston Carpenter, married Hannah, daughter of Benjamin and Sarah W. Acton. Heirs: John Redman Carpenter (married Mary C., daughter of Joseph B. Thompson.) (Heirs: Preston and Elizabeth.) Sarah Wyatt (married Richard H. Reeves). (Heirs: Augustus, Hannah, Mary and Alice.) Samuel Preston, Jr. (married Rebecca Bassett). (Heir: Benjamin K.) William Samuel P. Carpenter, married secondly, Sarah Sheppard.

VIII.—Margaret Carpenter, daughter of Preston and Hannah Carpenter, married James M. Woodnutt, of Salem County, N. J.

X.—MARTHA CARPENTER, daughter of Preston and Hannah Carpenter, married Joseph Reeves, of Salem County, N. J. Heirs: Samuel, Milicent, Joseph and Mary. (Samuel married Achsah Stratten, Milicent married Jos. Owens.)

DESCENDANTS OF JOHN CARPENTER (SECOND SON OF SAMUEL CARPENTER, SEN.)

JOHN CARPENTER married Anne, daughter of Richard and Esther Hoskins, 11 mo. 11, 1710. Richard Hoskins was "an eminent physician and minister of the Gospel." He died in England on a visit, about 1700. His wife died in Philadelphia, in 1698. He left several daughters.

Martha Carpenter, daughter of John and Anne, married in Philadelphia, Reese Meredith, March 23, 1738. Reese was the son of Reese, of Radnorshire, Wales. He produced a certificate dated 2d mo., 1730, from the monthly meeting in Leominster, Hereford County, Great Britain, of his right of membership among Friends. Died in Philadelphia, November 17, 1778, aged about 70. His wife died 8 mo. 26, 1769. He was a shipping merchant largely in trade. Their children were: Samuel, married Margaret Cadwallader. Anne, married Henry Hill, merchant. Elizabeth, married George Clymer. George Clymer was born in Philadelphia, June 10, 1739.

ADDENDA TO THE CARPENTER FAMILY.

According to J. Smith's collection of Memorials, "Samuel Carpenter came to Philadelphia from Barbadoes in 1683. The tradition among the oldest of his descendants is that he came from England. It is therefore probable that he was a native of that country, but had resided awhile in Barbadoes, for the purposes of trade. Of his parentage and early history nothing is now known. His papers appear to have been entirely lost, and the few facts relating to his life which are here collected have been gleaned from the MSS. of his contemporaries.

John R. Carpenter, in his MSS. collection, says:

"The following particulars, few and imperfect, are all that I have been able to obtain, after much research and inquiry, of the life of my ancestor Samuel Carpenter."

The "Globe Tavern," owned by Samuel Carpenter, was in aftertimes called "Peg Mullins' Beef Steak House." It was on the west side of Water Street, corner of Wilcox's Alley. The late aged Col. Morris says it was the fashionable house of his youthful days. Gov. Hamilton and other Governors held their Clubs here, and here the Freemasons met, and most of the public parties and societies.

Samuel Carpenter (2d) was born 9th February, 1688, and died Nov., 1748, aged 60 years.

Preston Carpenter, second son of Samuel Carpenter 2d, married Hannah, daughter of Samuel and Hannah Smith, and granddaughter of John Smith, of Hedgefield, Salem County, New Jersey.

In 1693 Samuel Carpenter was one of the members of the Provincial Assembly, but when elected, or how long he served, is uncertain.

In 1697 he was one of the Governor's Council of State, and continued to serve in that capacity until his death, 1714. At that time, and for several years previous, he was Treasurer of the Province. Gov. Penn and his deputies for many years sustained a violent and able opposition from a numerous party in the Provincial Assembly, who demanded a more democratic form of Government than the Executive thought proper to allow.

JOSHUA CARPENTER.

"So far as I have been able to ascertain, Samuel Carpenter was accompanied by but one of his family—a brother named Joshua— when he came to Pennsylvania. Joshua was a member of the Church of England. His wife's name was Elizabeth. They resided in Philadelphia, and left one or more children.

Some of his descendants resided near Dover, Delaware, one of whom had in his possession several old family portraits which had belonged to Joshua Carpenter. He (Joshua) built that ancient house which until recently stood in the rear of Judge Tilghman's residence on Chestnut Street, and was known as Græme Hall. It was intended as Joshua Carpenter's summer residence. The Arcade was subsequently erected on the site of this mansion.

It is claimed by the Carpenters of Elsenboro, Salem Co., N. J., that William Carpenter, grandson of Joshua, removed to Salem County about the year 1750, where he married Mary, daughter of Jeremiah Powell, who left four children—Mary, William, Powell and Abigail. Mary married Job Ware; William, Elizabeth Ware; Powell, Eliza Slaughter (and secondly, her sister, Ann Slaughter); Abigail married Edward Hancock. Powell was wounded at the massacre by the British, at Hancock's Bridge, Salem County, during the Revolution of 1776. William Carpenter, the head of this branch, was an Episcopalian, and buried in St. John's Episcopal Churchyard, Salem.

ABRAHAM CARPENTER.

The following is a copy of a letter written by J. E. Carpenter, 710 Walnut Street, Philadelphia, May 2d, 1879, to Charles Perrin Smith, Esq.:

TRENTON, N. J.

MY DEAR SIR.—I have just found some valuable material to assist in locating the family of our ancestor, Samuel Carpenter, in the old country. I had known before that Samuel and Joshua Carpenter had a brother Abraham, who was with them here in Philadelphia, but whether he was permanently a resident of Philadelphia, or whether he died here, I did not know. Recently, in making some investigations in the office of Register of Wills, of this city, I found the *Will of Abraham Carpenter.* Most fortunately, it gives more information respecting the members of the family in England than any authentic document in existence that I know of. The following are briefly some of its provisions:

It is called the Will of Abraham Carpenter, Merchant. He leaves all of his estate to his brother Samuel, and his brother Samuel's son Samuel, in trust to pay legacies, &c.

A legacy is left to his sister Mary (widow) in Lambeth, England.

A legacy to Damaris, wife of David Hunt, of the borough of Southwark, his sister.

A legacy to the children of his brother, John Carpenter, late of Horsham, in Sussex.

A legacy to his cousin (perhaps a grand-niece, as the term is used in another place in his Will in this sense) Susanna, daughter of John and Ann Welch, of Southwark. (Ann Welch was perhaps the daughter of his sister Damaris Hunt.)

A legacy to the children of his sister Deborah Jupp, deceased.

A legacy to the children of his sister Mary.

A legacy to the children of his sister Damaris.

A legacy to his kinsman, Thomas Mitchell.

A legacy to his cousin, Robert Story, to be paid when twenty-one years of age, or married.

Robert Story was the son of Enoch Story, who married Sarah, the daughter of Joshua Carpenter. He was therefore the son of Abraham Carpenter's niece, or what is now termed a grand-nephew.

A legacy to his cousin, Sarah Fishbourn, when 21 or married. (Sarah Fishbourn was probably the daughter of William Fishbourn, who married Hannah, daughter of Samuel Carpenter, Sen'r, and the same relationship appears a grand-niece, and evidently this is the case, because all of these cousins are minors.)

A legacy to Samuel, son of his brother Joshua.

A legacy to Samuel and John, sons of his brother Samuel.

A legacy to his brother Joshua.

A legacy to Hannah Hardiman. (She was probably a daughter of Benjamin Hardiman, brother of Hannah Hardiman, who married Samuel Carpenter, Sen'r.

A legacy to his sister-in-law Elizabeth, wife of his brother Joshua.

A legacy to Hannah, wife of his brother, Samuel Carpenter, Sen'r.

His negro woman, Hagar, to live with his brother Samuel and his wife, and her son Ishmael shall live with her; gives Ishmael to his brother Samuel; appoints his brother Samuel and his brother Samuel's son, Samuel, Executors. Dated March 27, 1708. Registered Philadelphia Will Book, C, p. 87. Proved April 14, 1708.

You will observe that the relationship is proved in every particular, the names of brothers, brothers' wives and children correspond exactly, and the names of Hardiman, Fishbourn and Story, make it impossible that he could belong to any other family than our own; but having started this rich mine of information, I was curious to follow the clue, and so searched for and found the Will of Joshua Carpenter.

It is called the Will of Joshua Carpenter, Brewer, and makes the following provisions:

A legacy to his grandson, Robert Story, the son of his daughter Sarah, to be paid him at twenty-one years of age.

A legacy to his grand-daughter, Patience Story, daughter of his daughter Sarah, to be paid her at eighteen, or at her marriage.

A legacy of one shilling to his son-in-law, Enoch Story.

A legacy to each of his sisters Mary and Damaris, near London, in Great Britain, to be remitted to them, if they then be living.

A legacy to his cousin, Anne Busfil, and to her children, William, Mary, Mercy and Joshua Busfil, each to be paid when sons are 21, and daughters 18, or married.

A legacy to his cousins Abraham and Thomas Mitchell, and residue to his wife, Elizabeth.

Dated Aug. 27, 1720. Proved Aug. 2, 1722. Registered at Philadelphia in Book of Wills, D, p. 325.

You have probably seen the Will of Samuel Carpenter, the elder. It mentions his brother Joshua, but makes no mention of Abraham, who died before him. In fact, his entire estate being divided among his widow and children, he makes no mention whatever of his collateral relatives, and only mentions his brother Joshua's name as owning a lot adjoining his property, in describing the boundaries of the property devised.

Yours very truly,   (Signed) J. E. CARPENTER.

Descendants of Thomas Carpenter, of Carpenter's Landing, N. J., (son of Preston, whose only son Edward married Sarah Stratton.)

DESCENT.—Samuel (1), Samuel (2), Preston, Thomas, Edward.

Edward Carpenter, son of Thomas Carpenter, of Carpenter Landing, N. J., and a lineal descendant of the fifth generation of Samuel Carpenter, Sen'r, of Philadelphia. Treasurer of the Province of Pennsylvania, Member of the Provincial Assembly, Commissioner of Gov. Penn during his absence in England, etc. Married Sarah, daughter of James Stratton, M.D, of Swedesborough, Gloucester, County, New Jersey.

Thomas Carpenter, the father of Edward above-named, although born and educated a Quaker, served as an officer in the New Jersey line during the Revolution of 1776, holding the rank of Paymast'r and Commissary. He was present at the battles of Assanpink a: 1 Princeton, and the companion of Gen'l Mercer the night before the battle in which the latter was killed. (MSS. of J. E. C.) He was at Red Bank immediately after the close of the battle, and aided in taking care of the wounded Hessian General Count Dunop.

Washington was greatly indebted to Commissary Carpenter for the subsistence of his army while in winter quarters at Morristown, New Jersey. Provisions and forage were chiefly drawn from South Jersey on sledges during that eventful period by Commissary Carpenter, escorted by a body of dragoons. He collected the supplies mostly from the counties of Gloucester, Salem and Cumberland. (MSS. J. E. C.)

Edward Carpenter married Sarah Stratton Sept. 5, 1799. He resided at Glassboro, New Jersey, where he succeeded to his father's interest in the Glass Works, now (1879) owned by T. H. Whitney & Bros., which he retained until his death, March 13, 1813. His widow survived until February 12, 1852. Upon her devolved the task of educating her five children. She inherited the fine qualities of her father, Dr. James Stratton. She was a strict disciplinarian, an exemplary Christian, and possessed great strength of character.

After her husband's demise, she removed with her family to Woodbury, and subsequently to Carpenter's Landing, where she remained at the head of her father-in-law's house until he died, in 1847, when the mansion was sold.

The remainder of her life was divided among her children. She died at the residence of her son, Edward Carpenter, in Philadelphia, and was buried with her husband in the old graveyard, Trinity Church, Swedesborough. Their children were, 1. Thomas Preston Carpenter, born April 19, 1804; married, Nov. 19, 1839, Rebecca Hopkins, daughter of Samuel Hopkins, M.D.

He received a liberal education and was admitted to the Bar of New Jersey, where he attained deservedly high eminence.

He was appointed Justice of the Supreme Court of New Jersey during Geo. Stratton's official term. At the close of his official term, Judge Carpenter removed to Camden, where he continued in the practice of his profession until his demise, March 2d, 1876. Judge Carpenter was a prominent member of the Episcopal Church, and for many years represented the Diocese in which he lived as a Delegate to the General Convention. He was noted for many accomplishments, genial manners and pleasing address. His wife survives him (1879). Their children,

1. Susan Mary Carpenter, born Aug. 4, 1840.
2. Anna Stratton Carpenter, born June 10th, 1843; died Dec. 13, 1869.
3. Thomas Preston Carpenter, born Sept. 23, 1846; died Aug. 25, 1848.
4. James Hopkins Carpenter, born Nov. 18, 1849.

2. Mary Tonkin Carpenter, second child of Edward and Sarah Carpenter, born Sept. 14, 1805; married, March 24, 1830, Richard W. Howell, of Camden, N. J., a lawyer of ability and eminence. He was the son of Col. —— Howell, of "Fancy Hill." Richard W. Howell died Aug. 12th, 1859, and was interred in the cemetery at Camden. Mrs. Howell still survives (1879). Her children are:

1. John Pascall Howell, born 1831, died 1842.
2. Edward Carpenter Howell, born 1833, died 1834.
3. Samuel Bedell Howell, born 1834; married Maria Neill and has several children.
4. Charles Stratton Howell, born 1837 (unmarried).
5. Richard H. O. Howell, born 1840, died 1850.
6. Joshua Ladd Howell, born 1842; married Mary E. Savage, and has a child.
7. Thomas James Howell, born Oct. 10, 1844; killed in battle of "Gaines Mills," Virginia, June 27, 1862. He was Lieutenant in Company I; went through all the fight uninjured, and acted with great bravery. But after his regiment came out of the woods and was forming to cross the bridge, he was struck by a chance and nearly spent cannon ball, which went clear through him, killing him

instantly. The Lieutenant was in his eighteenth year, and a talented and worthy young man. (New Jersey and the Rebellion.)

8. Anna Howell, born Sept. 12, 1846; married Malcom Lloyd, Esq., and has children.

9. Francis Lee Howell, born 1849, died 1872.

10. Sarah Carpenter Howell, born 1850, died 1850.

3. James Stratton Carpenter, third child of Edward and Sarah Carpenter, born October 14, 1807; married, Oct. 12, 1832, Camilla Julia Sanderson, daughter of John Sanderson, Esq.

He graduated at the University of Penna., where he took the degree of M.D. He visited Europe and continued his medical education in Paris.

Returning home in 1839, he established himself at Pottsville, Penna., then but recently settled, where he continued in the practice of his profession until his decease, Jan. 31, 1872.

His reputation as a skillful physician and surgeon was not confined to his immediate neighborhood, but he was frequently called for consultation to distant sections of the country. His social qualities and great hospitality created for him many warm friendships. He was interred at Pottsville. His widow yet survives (1879). Their children:

1. John Thomas Carpenter, born 1833; married, 1855, Eliza Hill, daughter of Charles M. Hill, Esq., of Pottsville. They have four children.

He graduated L.B. with first honor, at University of Penna., 1852; took the degree of A.M. in the same University, 1855. He also graduated M.D., Medical Department, same University, in 1852. He served with great distinction in the Union Army, having been appointed Surgeon by the Governor of Pennsylvania at the outbreak of the Rebellion, April 30, 1861. He was attached to the 34th Regt. of Infantry, Penna. Reserves, June 6th, 1861; Medical Director of Gen. McCook's Brigade, Army of West Virginia, Oct. 14, 1861; Medical Director in charge of General Hospitals, Cumberland, Maryland, March, 1862; Medical Director of Mountain Department, Wheeling, Va., May 10, 1862; in charge of General Hospitals, Cincinnati, Ohio, Aug. 25, 1862: Medical Director, Department of Ohio, Dec., 1863; Medical Director and Superintendent of Hospitals, District of Columbia, March 19, 1864; President of Army Medical Board, Cincinnati, May, 1863.

He resides (1879) at Pottsville.

2. Sarah Stratton Carpenter, born June 14, 1835; married Rev. Daniel Washburn, Jan. 27, 1854. He is now Rector of Episcopal Church, Ashland, Pa., and has children.

3. Sophia Carre Carpenter, born Nov. 11, 1837.

4. Caroline Maria Carpenter, born Dec. 18, 1840.

5. James Edward Carpenter, born Sept. 29, 1843, died Jan., 1845.

6. Preston Carpenter, born Sept. 29, 1843; married Kate, daughter of Edward Wheeler, of Pottsville. He had two children. He served in the Union Army during a portion of the war of the Rebellion, attached to the Signal Corps. He married, secondly, a sister of his first wife, the widow of William Parry, late of Pottsville.

7. Camilla Carpenter, born June 10, 1851.
8. Mary Howell Carpenter, born Nov. 17, 1856.
9. Richard Howell Carpenter, born March 2, 1858.

4th. Samuel Tonkin Carpenter, fourth child of Edward and Sarah Carpenter, born Nov. 28, 1810; married, May, 26, 1841, Francis Champlain, of Derby, Connecticut, who died Jan'y 4, 1845. Married, secondly, Emily Thompson, of Wilmington, Delaware. He was a clergyman of the Episcopal Church, and settled for some years in Connecticut. Afterward he became rector of the Episcopal Church at Smyrna, Delaware, where he remained many years.

He also had other charges. Subsequently he became chaplain in the army. He died Dec. 26, 1864, and was interred in the new graveyard, Trinity Church, Swedesborough.

His second wife survives him and resides at Joliet, Illinois. Children by his first wife:

1. Samuel Champlain Blakeslee Carpenter, born Nov. 10, 1842; died Sept. 28, 1871. He served with credit in the Union Army during the war of the Rebellion.
2. Frances Mary Carpenter, born July 21, 1844. Children by second wife:
1. Herbert Dewey Carpenter, born June 2, 1853.
2. Florence Carpenter, born June 2, 1853.
3. Horace Thompson Carpenter, born Oct. 10, 1852.
4. Richard Howell Carpenter, born Dec. 21, 1861.
5. Lewis Tonkin Chatfield Carpenter, born Nov. 17, 1864.

5. Edward Carpenter, fifth child of Edward and Sarah Carpenter, born May 17, 1813; married, Nov. 16, 1837, Anna M., daughter of Benjamin M. Howey, of "Pleasant Meadows," Gloucester County, N. J. After his marriage he resided a short time in Glassborough, N. J., and subsequently for a few years in Kent County, Maryland. He removed to Philadelphia, in 1843, where (with a short interval) he has since resided. He studied law, but turned his attention more particularly to the real estate branch of the profession.

He enjoyed an excellent reputation as a scientific conveyancer, and his work bears a professional value second to none in Philadelphia.

He is a prominent member of the Episcopal Church, and was one of the founders of the Church of the Mediator, in Philadelphia. He and his wife both survive. Their children are:

1. Lewis Henry Carpenter, born Feb. 11, 1839 (unmarried). Captain 10th U. S. Cavalry, Brevet-Col. U. S. Army, now (1879) on duty in Texas. He served in the cavalry of the Army of the Potomac, during the war for the Union, entering the service as private in the 6th U. S. Cavalry. He received every Brevet from 1st Lieutenant to Colonel for gallant and meritorious conduct.

Before the close of the war he commanded a regiment with the rank of Colonel of Volunteers. He served on the personal staffs of Generals Pleasanton and Sheridan, and participated in all the great cavalry battles of Virginia.

Since the close of the war he has been stationed in the Indian country, and had the honor of being mentioned several times in the reports of the senior officers under whom he served.

In one instance attention was called to his brilliant conduct in the battle of Beaver Creek, Kansas, by special order from Department Headquarters, issued by Lieutenant-General Sheridan. For this action he received his Brevet as Colonel.

2. James Edward Carpenter, born March 6th, 1841; married, Oct. 17, 1867, Harriet Odin Dorr, daughter of Rev. Benjamin Dorr, D.D., rector of Christ Church, Philadelphia.

He served as a volunteer in the war of the Rebellion, enlisting as a private in the Eighth Pennsylvania Cavalry. He became Captain and Brevet-Major. During the latter part of the war he served on the staff of Major-General Gregg (one of Sheridan's Division Commanders), and participated in nearly all the battles fought by the Army of the Potomac. He was wounded in the cavalry fight of Philamont, during the advance of the army into Virginia after Antietam. In the charge of the Eighth Pennsylvania Cavalry at Chancellorsville, his horse was shot under him, and of the five officers who rode into the fight, he was one of the two who survived. Having partially prepared for the bar prior to the commencement of the war, he resumed its studies at its close, and in October, 1865, was admitted at Philadelphia in the full practice of his profession.

He is (1877) a vestryman of Christ Church, Philadelphia, a manager of Christ Church Hospital and of Christ Church Chapel, Treasurer of the Historical Society of Pennsylvania, and a member of the Executive Committee of that body. He has two children living.

1. Edward, born Aug. 27, 1872.
2. Helen, born Nov. 11, 1874.
Grace, born Oct. 25, 1876; died May 27, 1877.

3. Sarah Caroline Carpenter, born Jan. 18, 1843; married, Jan. 8, 1855, Andrew Wheeler, a prominent iron merchant, and vestryman of St. Luke's Episcopal Church, Philadelphia. Their children are:

Andrew, Samuel, Bowman, Arthur, Leslie, Walter, Stratton and Herbert.

4. Mary Howell Carpenter, born Jan. 22, 1845.

5. Caspar Wistar Carpenter, born Sept. 17, 1847; died 1848.

6. Thomas Preston Carpenter, born April 30, 1849. Resides in Chicago (unmarried, 1879).

7. Henrietta Howell Carpenter, born 1855; died young.

8. Charles Creighton Carpenter, born Dec. 11, 1860; now (1879) student in Pennsylvania University.

[From the records compiled by the late C. P. Smith, through the courtesy of his daughter Miss Elizabeth A. Smith, of Trenton, New Jersey].

CARPENTER FAMILY OF MASSACHUSETTS.

1. William Carpenter (1), the ancestor, it is supposed of all the Carpenter family in New England, was born in England, 1576, and left Harwell in 1638, and went on board the ship *Bevis*, at Southampton, and arrived the same year (aged 62), and stopped at Weymouth. With him came his son William (2), aged 33, and Abigail, wife of the latter, aged 32, and four children, " of ten years old, or less," who are not named in the clearance of the vessel. From other sources we learn that the names of these children were, William (2), Samuel (2), Joseph (2) and John (2). He resided some time at Weymouth. He died in the winter of 1659-60. His descendants are very numerous.

William Carpenter (2), son of the preceding, born in England, 1605, came with his wife Abigail—born in England 1606, and four children, as already stated. He removed with Rev. Samuel Newman and a majority of his church, from Weymouth to Rehoboth (the part now known as Seekonk), and with them began the settlement of that place in 1645. He was town clerk of Rehoboth from 1645 to 1649. He was deputy to Plymouth General Court in 1656. After the death of his father, I suppose, he went back to Weymouth, where he had first resided. He must have died previous to May 26, 1667, since, in a division of meadow land then made in the North Purchase (now Attleboro' on Cumberland), "Widow Carpenter" is mentioned. She died February 22, 1688. His children were:

William (3), b. 1631; m. 1, Priscilla Bonett, Oct. 5, 1651. She died Oct. 20, 1663. M. 2, Miriam Searle, Dec. 10, 1663. She died May 1, 1722, in Rehoboth, a. 76.

Samuel (3), b. ——; m. Sarah Readway, May 25, 1660. He died Feb. 20, 1682. She, for her 2d husband, married Gilbert Brooks, all of Rehoboth.

Joseph (3), b. ——; m. Margaret Sabin, Nov. 25, 1655. He died May 6, 1675. She died 1700, aged 65, at Swanzey, Mass.

Their children were: Joseph, b. Aug. 15, 1656. Benjamin, Jan. 19, 1658. Abigail, March 15, 1659. Esther, March 6, 1661. Martha, 1662. John, Hannah, Jan. 21, 1672. Solomon, April 27, 1673; d. next year. And Margaret, May 4, 1675.

John (3), b. ——; m. Hannah, ——; was in the "Narragansett Expedition," 1675; resided at Jamaica, L. I.; died May 23, 1695.

His children by wife Hannah were: Amos, b. Nov. 19, 1677. Eliphalet, April 17, 1679, and perhaps by his wife Dorothy — m. Feb. 9, 1680; had Priscilla, Jan. 20, 1681.

Abiah (3), b. ——; went to Warwick, R. I., to a tract of land bought by his father.

Hannah (3), b. at Weymouth, April 3, 1640.

Abraham (3), b. at Weymouth, April 9, 1643.
(In another record this is given as Abia; daughter same date.)
(Geneal. Reg., Oct., 1854.) *O----,a--* *A\ /.* *?-.~* *--.?.*
Ephraim (3), b. at Rehoboth, April 25, 1651; 'd. April 30, 1713,
a. 62.

THIRD GENERATION.

*William Carpenter (3), son of the preceding, b. in England, 1631.
Married 1st, Oct. 5, 1651, Priscilla Bonett; 2d m., Miriam Searle,
December 10, 1664. He was town clerk of Rehoboth from 1668 to
1703, except 1693. He was often employed in town business. He
d. Jan. 26, 1703. His children were by first wife Priscilla:
John (4), b. Oct. 19, 1652; m. Rebecca ——; settled at Wood-
stock, Conn.
William (4), b. June 20, 1659; m. Elizabeth Robinson, April 8,
1685; d. in Attleboro', March 10, 1719.
Priscilla (4), b. July 24, 1661; m. Richard Sweet.
Benjamin (4), b. Oct. 20, 1663; m. Hannah Strong. He moved
away before 1689, and d. at Coventry, Conn., April 18, 1738. She
died March 20, 1762, a. 92.
Son of William by his second wife:
Josiah (4), b. Dec. 18, 1664; m. Elizabeth Read, Nov. 24, 1692.
He d. Feb. 28, 1727; and she d. Oct. 18, 1730, a. 72, in Attleboro'.
Nathaniel (4), b. May 12, 1667; m. Rachel Cooper, Sept. 19, 1693.
She d. July 9, 1694, a. 23. He m. 2d, Mary Preston, Nov. 17, 1695.
She d. May 25, 1706, a. 31. He m. 3d, Mary Cooper, July 8, 1707.
She d. April 9, 1712, a. 30. He then m. Mary Bacon, 1716.
Daniel (4), b. Oct. 8, 1669; m. Bethiah Bliss, April 15, 1695. She
d. February 27, 1703, a. 31. He then married Elizabeth Butter-
worth March 30, 1704. She d. June 13, 1708, a. 26. He then m.
Margaret Hunt, March 19, 1718. She d. 1720. He then m. Mary
Hyde; he died Sept. 14, 1721.
Noah (4), of Attleboro, b. March 28, 1672; m. Sarah Johnson, Dec.
3, 1700; had Noah, born Nov. 25, 1701; Marian, Dec. 25, 1702; Sarah,
Sept. 24, 1704; Stephen, July 23, 1706; Asa, March 10, 1708; Mary,
Jan. 24, 1710; Margaret, March 30, 1712; Simon, Nov. 13, 1713; d. next
m.; Isaiah, Feb. 7, 1715; Simon (again) Aug. 20, 1716; Martha, May
25, 1719; Elisha, Aug. 28, 1721; Amy, Feb. 2, 1723.
His wife died Sept. 29, 1726. He m. secondly, May 22, 1727, Ruth
Follet, and had Priscilla, May 1, 1728. This wife died June 10,
1745. Next he m. Tabitha Bishop, 174—; he d. June 7, 1753, in
Attleboro'.
Mirriam (4), b. Oct. 26, 1674; m. Jonathan Bliss, June 23, 1691.
Obadiah (4), b. March 12, 1678; m. Deliverance Preston, Nov. 6,
1703. She d. June 12, 1767, a. 85. He d. at Rehoboth, October 25,
1749.
Ephraim (4), b. April 25, 1681; m. Hannah Read, Aug. 14, 1704.
She d. August, 1777, a. 36. He then m. March 24, 1718, widow Mar-
tha Carpenter. He d. at Rehoboth, April 20, 1745.

---

* He was granted a coat-of-arms, May 4, 1683.

Hannah (4), b. April 10, 1684; m. Jonathan Chase, Nov. 23, 1703.
Abigail, b. April 15, 1687; m. Daniel Perrin, Nov. 12, 1706. Daniel Perrin was town clerk from 1668 to his death. His widow died May 1 or 7, 1722.

FOURTH GENERATION.

Samuel (3), brother of William Carpenter (3), b. in England, not far from 1633; m. Sarah Readway, May 25, 1660. Lived in Rehoboth. He advanced money to carry on "Phillip's War." He d. February 20, 1682. Afterwards, his widow, Sarah, m. Brooks. The children of Samuel and Sarah were: Samuel (4), born Sept. 15, 1661; Sarah (4), b. Jan. 11, 1664; Abiah (4), b. Feb. 10, 1666, d. 1732; James (4), b. April 12, 1668, d. 1738; Jacob (4), b. Sept. 5, 1670; Jonathan (4), b. Dec. 11, 1672, d. 1717; Solomon (4), Dec. 23, 1677; David (4), April 17, 1675; Zechariah (4), July 1, 1680; Abraham (4), b. Sept. 20, 1682.

Joseph. Hon. Abbot Lawrence and Hon. Benjamin Carpenter were descendants of Joseph. The following inscription is from the tombstone of Hon. Benjamin Carpenter's grave in the west part of Guilford, Vt.: "Sacred to the Memory of Hon. Benjamin Carpenter, Esq., a Magistrate in Rhode Island in A. D. 1764, a public teacher of righteousness; an able advocate to his last of democracy and equal rights of man; removed to this town A. D. 1770; was a field officer in the Revolutionary War; a founder of the first Constitution and government of Vermont; a Councilor of Censors in A. D. 1783; a member of the Council and Lieutenant-Governor of the State in A. D. 1779; a firm professor of Christianity in the Baptist Church for 50 years. Died March 29, 1804, aged 78 years, 10 months and 12 days."

Josiah (4), son of William (3) and Miriam; m. November 24, 1692, Elizabeth Reed; b, 1667. He was a cordwainer; lived in Attleboro', and died there Feb. 28, 1727, a. 64. She died at Attleboro', Oct. 18, 1739, a. 72. Their children were:
Josiah (5), b. March 4, 1693–4; d. May 18, 1716, a. 16.
Edward (5), b. April 23, 1695; d. June, 1696.
Seth (5), b. Dec. 5, 1697; d. Jan. 24, 1698–9.
Elizabeth (5), b. March 19, 1699–1700; m. Israel Peck, 1727.
Comfort (5), b. May 8, 1709; m. Huldah Bowen, Nov. 12, 1730.

Benjamin, of Northampton, Mass., son of William, of Rehoboth, Mass.; born Oct. 20, 1663; m. March 4, 1691, Hannah, daughter of Jedediah and Freedom (Woodward) Strong, who was b. Feb. 3, 1671, in Northampton, Mass. He was a farmer in Northampton, Mass., and after 1708 in Coventry, Conn. He d. in Coventry, April 18, 1738. She d. March 20, 1762. Their children were:
Freedom, b. July 13, 1692; (m. A. B. Carpenter, gives also Prudence and same date). Amos, b. Nov. 6, 1693. Benjamin, b. Oct. 3, 1695. Jedediah, b. Oct. 1, 1697. Hannah, b. Aug. 15, 1699. Eliphalet, b. Oct. 16, 1701, d. Aug. 28, 1702. Eliphalet, b. Nov. 29, 1703. Noah, b. Dec. 24, 1705. Elizabeth, b. June 15, 1707. Ebenezer, b. Nov. 9, 1709. Rebecca, b. Nov. 23, 1711.
Noah (4). His children were:

Noah, b. Nov. 25, 1701; m. Persis Follett, June 6, 1728. He d. June 7, 1753. She d. 1753.

Eliphalet Carpenter, son of Benjamin and Hannah (Strong) Carpenter, was born Nov. 29, 1703, in Northampton, Mass., and shortly after his birth his parents removed to Coventry, Conn. He married, Nov. 1, 1727, Elizabeth, daughter of John and Hannah (Gillett) Andrews, born in Hartford, Ct., Feb. 17, 1705-6. She died May 6, 1773, and he married, second, Oct. 26, 1775, Abigail Ladd. She died —— He, a farmer in Coventry, Conn., where he died Feb. 22, 1792. He held a Captain's commission from the King. Children:
* Hannah, b. March 22, 1728; d. June 5, 1740.
Elizabeth, b. April 15, 1731; m. Ephraim Root, d. Dec. 30, 1751.
* Asahel, b. Jan. 30, 1733; d. June 9, 1740.
* Lois, b. May 21, 1735; d. June 4, 1740.
* Abigail, b. Dec. 9, 1736; d. June 8, 1740.
* Anna, b. April 9, 1739; d. June 7, 1740.
Hannah, b. May 15, 1741; d. Oct. 8, 1742.
Submit, b. Jan. 27, 1743; m. Nov., 1764, Reuben Stiles; d. Dec. 26, 1837.
Lois, b. Dec. 13, 1745; m. Isaiah Porter; d. April 4, 1766.
2. Eliphalet, b. Nov. 9, 1747; m. ——
2. Eliphalet C., b. Nov. 9, 1747, in Coventry, Conn. Married, May 22, 1766, Esther, daughter of Jonathan and Hannah (Baker) Gurley. She was b. in Mansfield, Conn., June 16, 1749, and d. Oct. 23, 1819. He, a farmer in Coventry, Conn., where he died Dec. 21, 1820. Children :
Lois, b. Nov. 8, 1768; d. Sept. 9, 1770.
Artemas, b. Sept. 21, 1770; died March 3, 1837, at German Flats, N. Y.
Anna, b. Sept. 15, 1772; m. Solomon Judd; d. Jan. 29, 1847.
Esther, b. March 25, 1775; m. Thomas Judd; d. Jan. 10, 1846.
3. Ralph, b. Aug. 2, 1777; m. ——
Hannah, b. March 19, 1780; d. April 3, 1803; unmarried.
Cynthia, b. Dec. 3, 1783; m. 1, Ezra Warner; 2, Aduch Abbott; d. Aug. 24, 1839.
Achrah, b. April 25, 1786; m. Samuel Topliff.
Lucy, b. Nov. 20, 1789; m. Elijah Dixter; d. April 21, 1831.
3. Ralph C., b. Aug. 2, 1777, in Coventry, Ct.; married, Dec. 27, 1801, Mary, daughter of Levi and Hannah (Draper) Spicer. She b. —— and died May 30, 1858. He, a farmer in Coventry, where he d. April 2, 1850. Children:
William Riley, b. Jan. 1, 1802; d. Sept. 9, 1808.
Hannah Spicer, b. Feb. 16, 1804; living unmarried in Coventry, Ct.
4. Anson, b. Dec. 20, 1806; m. —— ; d. ——
Maria, b. March 12, 1809; m. May 25, 1847, Nathan Merrow, Oct. 24, 1867.
Caroline, b. Sept. 11, 1813; m. Sept., 1837, Joshua Tilden; d. Oct. 6, 1843.

---

*Died of Black Throat Distemper within a few days of each other.

5. William Riley, b. April 19, 1816; m. ——
Lucy, b. July 19, 1818; d. April 21, 1831.
Lydia, b. Oct. 15, 1820; m. Sept. 26, 1847, John H. Palmes; resides at Hadlyme, Ct.
6. Ralph Monroe, b. Dec. 7, 1822; m. 1st, Sarah G. Root, 2d, Mrs. Nancy Clark.
Mary Ann, b. April 30, 1826; d. Nov. 23, 1844.
4. Anson C., b. Dec. 20, 1806, in Coventry, Conn; married Feb. 15, 1832, Diantha, daughter of Warren A. and Anna (Day) Skinner. She b. in Colchester, Ct., Sept. 19, 1811, and died Nov. 9, 1873. He, a farmer in East Hampton, a parish in Chatham, Conn., where he died Sept. 9, 1856. Was a member of Conn. Legislature, sessions of 1850 and 1851. Children:
Eldredge Spicer, b. Jan. 23, 1833; d. May 24, 1857; unmarried; preparing for the ministry.
7. Don Carlos, b. Nov. 19, 1834; m. ——; d. ——
Lucy Ellen, b. Feb. 15, 1837; m. Feb. 21, 1858, Jane R. Clark, Feb. 4, 1880.
8. Hubert Edgar, b. March 19, 1839; m.——
9. Legrand Stiles, b. Oct. 14, 1841; m. ——
10. Gwinnett, b. Jan. 31, 1844; m. ——
Caroline Tildin, b. Dec. 24, 1845; m. May 5, 1867, William P. Waite; 2d, Geo. F. Jones, of Newton, Kansas.
Mary, b. Feb. 17, 1848; m. May 23, 1881, George R. Landers.
Ralph Ambrose, b. April 14, 1851; d. March 5, 1864.
Ruth Ann, b. Sept. 3, 1853; m. March 3, 1875, Martin L. Roberts, of New Haven, Conn.
5. William Riley C., b. April 19, 1816; m. Feb. 25, 1840, Laura (Crowley), widow of Cyrus Goff. She d. May 4, 1875. He was killed by falling from a building in New London, Conn., where he resided, Dec. 28, 1878. Occupation, joiner. Child:
Lydia M., b. Dec. 2, 1840; d. Nov. 16, 1859.
6. Ralph Monroe C., b. Dec. 27, 1822; m. 1st, Feb. 27, 1847, Sarah J. Root, she d., and he m. 2d, Jan. 30, 1882, Mrs. Nancy Clark. A farmer in Coventry, Conn. Children:
Cynthia J., b. Nov. 9, 1847; m. May 28, 1873, Henry Walker.
Sherman R., b. Oct. 23, 1849; m.
Candace, b. Jan. 20,1853; m. May 9, 1877, William Buell.
Prudence, b. Feb. 17, 1855; m. Edgar Gorton, April 7, 1873.
Calvin, b. May 18, 1856; m. Aug. 31, 1882, Emma Chapman.
Mary Spicer, b. Aug. 3, 1858.
Sarah Belle, b. April 30, 1860; d. March 27, 1874.
Mittie Anna, b. April 12, 1861; m. March 12, 1882, Herbert Walker.
7. Don Carlos C., b. Nov. 19, 1834; m. Dec. 2, 1855, Alice Ann, daughter of Alvah and Susan (Gilbert) West. She b. April 25, 1836. He a mechanic, and a member of Co. C, 24th Regt. Conn. Vols., 1861–1865. Residence, East Hampton, Conn., where he d. Dec. 5, 1880. Children:
Lillie Belinda, b. Aug. 1, 1856.
Henry Ansen, b. May 22, 1858; d. Dec. 30, 1870.

Clara Antoinette, b. Jan. 8, 1860; m. Frank L. Griffith.
Susan Diantha, b. April 14, 1866.
Ralph Eugene, b. Feb. 24, 1868.
Alice May, b. Aug. 12, 1871.
Sherman Francis, b. April 25, 1874.
8. Hubert Edgar C., b. March 19, 1839; m. Oct. 1, 1865, Anna,
daughter of John and Ann (Adamson) Hodge, b. in Edinburgh,
Scotland, July 23, 1844. He a member of Co. H, 21st Regt. Conn.
Vols., 1861-1865, and was severely wounded at the battle of Cold
Harbor, Va., June 3, 1864. Is a joiner, and was member of Conn.
Legislature, sessions 1878, 1879. Resides in East Hampton, Conn. ·
Children:
Eldridge Roscoe, b. Dec. 2, 1866; d. Nov. 4, 1877.
Anson Hodge, b. April 27, 1872; d. Oct. 29, 1877.
John Skinner, b. Oct. 13, 1882.
9. Legrand Stiles C., b. Oct. 14, 1841; m. Nov. 26, 1862, Ealenor
Melissa, daughter of Silas and Mary (Goff) Hills. She b. Feb. 16,
1843. He a butcher in East Hampton. Children:
Vivia, b. May 22, 1871; d. May 23, 1871.
Crayton Farnsworth, b. Aug. 30, 1872.
Milton Legrand, b. Jan. 22, 1874.
Howard Silas, b. May 17, 1877.
10. Gwinnett C., b. Jan. 31, 1844; m. Jan. 1, 1872, Lucy Elizabeth,
daughter of Henry S. and Emeline C. (Lord) Selden. She b. May
2, 1846. He a member of Co. H, 21st Regt. Conn. Vols., 1861-1865.
A farmer and mechanic in East Hampton, Conn. Children:
Amy Elva, b. Jan. 19, 1873.
Kirby Selden, b. Oct. 1, 1874.
Lucy Elizabeth, b. Dec. 27, 1881.
11. Sherman R. Carpenter, b. Oct. 23, 1849; m. May 18, 1879,
Anna M. Knight, of Whately, Mass. He is a farmer in Coventry.
Their child was:
Eveline B., b. Nov. 25, 1882.
Miriam (5), b. Dec. 25, 1702; d. March 1, 1726.
Sarah (5), b. Sept. 24, 1704; m. Noah Chase, May 5, 1720.
Stephen (5), b. July 23, 1706; m. Dorothy Whiticar, Nov. 28, 1724.
She d. Jan. 25, 1761.
Asa (5), b. March 10, 1708.
Mary (5), b. Jan. 24, 1709, at Rehoboth; m. John Read, April 19,   ✓
1733.
Margaret (5), b. March 30, 1712.
Simon (5), b. Nov. 13, 1713; d. Dec. 8, 1713.
Isaiah (5), b. Feb. 7, 1715; m. Widow Aletha Titus, Sept., 1734;
d. in Sutton, Mass.
Simon (5), b. Aug. 29, 1716; m. Sarah ——. He d. March 16,
1794, at Pomfret, Conn.
Martha (5), b. May 25, 1719.
Elisha (5), b. Aug. 28, 1721; m. Anne Whiticar, March 15, 1744.
He d. Aug. 2, 1789. She d. Feb. 23, 1804, at Sutton, Mass.
Anny (5), (Amy?) b. Feb. 2, 1723; d. Feb. 2, 1723.
Priscilla (5), b. May 1, 1728.

Abigail Carpenter (4), m. Nov. 12, 1706, Daniel (3) Perrin [John (2), John (1)], who was b. March 18, 1682. Their children were:
Abigail, b. Sept. 14, 1707; m. John Newman.
Susannah, b. Aug. 18, 1709.
Daniel, b. Feb. 10, 1710–11.
David, b. Oct. 15, 1714.
Mary, b. Jan. 11, 1716–7.
Noah, b. March 12, 1723–4.
Lydia, b. Jan. 17, 1726–7.
Hannah, b. Feb. 23, 1728–9.
Daniel Perrin, Jr., (Daniel, John, John), m. Sarah Hunt, April 8, 1736. Their children were:
David, b. March 25, 1737; d. young.
David, b. Oct. 20, 1739.
David Perrin, son of Daniel, Jr. (Daniel, Daniel, John, John); m. Abigail Cooper, April 29, 1762. Their children were:
Daniel, b. Feb. 15, 1763.
Susannah, b. Feb. 28, 1764; m. Thomas Carpenter, Dec. 24, 1788.
David, b. Oct. 10, 1765.
Thomas, b. March 1, 1768.
Noah, b. Feb. 23, 1770.
Abigail, b. Dec. 9, 1771.
Samuel, b. April 13, 1773; m. Orinda Walker, Feb. 13, 1800.
Ezra, b. Jan. 18, 1777.
Abigail, b. May 22, 1779.
Sarah, b. Aug. 3, 1781; m. Elijah Kent, Dec. 1, 1803.
Huldah, b. Aug. 6, 1783; m. Noah Cooper, June 7, 1808.
John, b. Feb. 6, 1786.
Daniel Perrin, son of the preceding, m. Esther ——. Their children were:
David, b. June 29, 1798.
Philena, b. Aug. 4, 1800.
Seba, b. May 21, 1802.
Daniel, b. May 25, 1804.
Nelson, b. April 13, 1809.
Mary, b. Sept. 14, 1811.
Thomas Perrin (son of David), b. 1768; m. ———. Their children were:
Otis, b. Feb. 18, 1791.
Asa, b. Nov. 12, 1792.
Thomas, b. Aug. 8, 1795.
Lewis, b. Aug. 7, 1797.
Lydia, b. July 24, 1800.
Polly, b. Nov. 4, 1802.
Amasa, b. March 5, 1805.
William, b. May 31, 1817.
Daniel Carpenter (4), son of William (3). About the 27th July, 1690, he fought in an engagement against Quebec. He is supposed to have held some position higher than a private, but whether in the Quebec engagement or later, is not known. As a townsman he was very popular, as is shown by the various offices he

was appointed to. His children by his first wife, Bethea, were, probably:

Daniel (5), b. Nov. 8, 1695; m. Susannah Lyon, of Woodstock, Vt., Dec. 29, 1720. "She was born at Woodstock, Sept. 29, 1699, and died at Rehoboth, July 7th, 1790, aged 90 years, 9 months and 2 days," as her gravestone, "In memory of Mrs. Susannah Carpenter, Widow of Daniel Carpenter, Esq.," reads. Daniel, d. Jan. 25, 1753. During his life he held town or other public office almost constantly.

Jabez (5), b. ——: m. for 2d wife, Betsy Monk. His children were: 1. Jabez (6), 2. Elizabeth (6), (or Sophie), m. James Read, 3. Heziah (6), m. Jacob Shorey. Their children were: Abel (7), Cynthia (7), Sally (7), (m. Noah Perry), Jacob (7), and Heziah (7). 4. Lucy (6), (m. James Cooper), children were: Samuel, Lucy (m. Atherton), and Betsey; 5. Bethial (6), (m. Aaron Lyon); children were: Obadiah, William, John, Betsey and Nabby; 6. Abigail (6), unm., Eleazer (5), b. ——. Children: Elihu, Abisha and Mary; she m. Peter Whitaker.

Lieut. Samuel Carpenter (4), eldest son of Samuel (3), and Sarah, b. in Rehoboth, Sept. 15, 1661; m. Jan. 8, 1683, Patience Ide, who d. Oct. 28, 1732, a. 68. He d. in Rehoboth. His children were: Samuel, b. Nov. 9, 1684, Timothy, Amos, Andrew, d. young, Patience, Andrew, Uriah, Josiah, Nathan, Charles, b. 1702; d. 1744; Edmund, Freelove.

FIFTH GENERATION.

Capt. Comfort Carpenter (5), son of Josiah (4), and Elizabeth, b. at Rehoboth, May 8, 1709; m. Huldah Bowen, Nov. 12, 1730. He graduated at Harvard College, 1730, and is said to have been a lawyer and a merchant in Rehoboth. Tradition reports he was killed by the Indians at Charleston, N. H., Sept. 13, 1739, a. 31. His children were:

Chloe (6), b. Aug. 20, 1731; d. Nov. 5, 1741, a. 11.
Cynthia, b. Sept. 21, 1733; m. Reynolds 2—Mayo.
Cyril, b. April 4, 1736; m. about 1759, Freelove Smith.
Orinda, b. March 13, 1738; m. Nathan Dresser, 1759.
Comfort, b. Jan. 26, 1739–40. (Posthumous.) M. —— Smith.

Samuel Carpenter, eldest of Lieut. Samuel (4), and Patience, b. in Rehoboth, Nov. 9, 1684; m. Hannah Johnson, Feb. 4, 1714. Removed to Pomfret, Conn. His children were: Samuel, Nathaniel, b. Nov. 20, 1718; m. 1st Mary Leffingwell; she d. July 9, 1764; 2d, Mary Durkee; Hannah.

Isaiah, son of Noah Carpenter, m. widow Alethea Titus. His children were (they removed to Sutton, Mass., about 1740):
Sarah, b. November 14, 1736.
Isaiah, b. September 27, 1738; d. November 1, 1748.
John, b. December 16, 1740; m. Hannah Record.
Jonah, b. October, 1744; m. Tervish Whitmore, November 22, 1769. He d. January 31, 1805. She d. August 29, 1834, in Ashford, Conn.

Daniel Carpenter (5), son of Daniel (4); m. Susannah Lyon. His children were:

Elisha (6), b. June 18, 1728.

Asabel (6), b. ——; m. Mary Shorey. Their children:
1, Susannah, m. Medbury; 2, Mary, m. Bliss; 3, Matilda, m. Tucker; 4, Sophia, unm.; 5, Asenath, unm.; 6, Christina Amelia, who m. Israel Droun; 7, Caroline Augusta, unm.; 8, Bethia, unm.; 9, Wooster, m. Lovina Brown.

Elizabeth (6), m. Atwood.

Susannah (6), m. Nathaniel Chaffeᵉ.

Hannah (6), unmarried.

Daniel (6), b. August 4, 1739; m. 1, Anna Lyon, by whom he had three children: 2d, married Olive Ide, by whom he had eight children. He d. April 18, 1823; buried in Providence Cemetery, and on his gravestone the family coat-of-arms is cut. His children were: 1, Elizabeth; 2. Abigail, m. James French; 3, Daniel, m. Rachel Lyon; 4, Drayton, m. Sally Peck; 5, Darivus, m. Anna Carpenter (daughter of Caleb Carpenter); 6, Betsy, m. Simeon Daggett. (Betsy, when 85 years of age, from memory painted a picture of her dead father, which her brother Davis now possesses); 7, Draper, m. Caroline Bassett; 8, Davis, b. March 25, 1794, m. Elpha French (she being descended from a family of Carpenters whose ancestor two generations before was Abiah Carpenter). Davis is the only surviving child of Daniel 3. All that is known of his family is the names of three of his children—Sarah F., Amelia, m. to Solon Carpenter (son of Wooster Carpenter), living in Providence, R. I., and Davis, Jr., living in St. Joseph, Mo. 9, Calvin, m. Abigail Tisdale; 10, Olive, m. Phanuel Jacobs.

SIXTH GENERATION.

Elisha Carpenter (6), b. June 18, 1728; m. Esther Greenwood, daughter of Revd. John Greenwood. Elisha was appointed ensign in that company whereof Phillip Walker, Esq., was captain, in a regiment of foot commanded by Colonel Thomas Doty, raised for a general invasion of Canada, 1758. It is likely he went to Savoy, to occupy a tract of land known as "Bullock's Grant," given him for his, or his father's, military services. He was one of the earliest settlers of Savoy, where he built the first saw-mill erected. He died there March 25th, 1813, and his wife, who was born May 4, 1733, died April 24, 1814. Their graves are unmarked by any monument to show their resting-place. Their children were born in Rehoboth, and are:

Esther (7) b. April 13, 1752; m. William Ingraham. She d. Savoy, July 26, 1846.

Cynthia (7), b. April 27, 1754; m. Nathaniel Braley. She d. Albion, N. Y., May 18, 1841.

Elisha (7), b. May 6, 1756; was a ship officer; lost at sea, 1785.

Benjamin (7), b. June 17, 1758; d. at Rehoboth, July 11, 1761.

Comfort (7), b. September 8, 1761; d. in southern port, yellow fever, August 8, 1785.

Hannah (7), b. March 17, 1764; m. Howland Kimball; d. Gaines, N. Y., September 10, 1828.

Sarah (7), b. Dec. 15, 1766; d. Rehoboth, Oct. 26, 1767.

Benjamin (7), b. Sept. 11, 1768; m. 1, Nancy Fisher; 2, Mima Hollis. He d. Savoy, June 11, 1836.

Sarah (7), b. Oct. 4, 1771; m. Pardon Arnold; d. Manchester, N. Y., July 2, 1857.

Nancy (7), b. Feby. 5, 1775; d. Rehoboth, Jany. 17, 1767.

Elijah (7), b. May 26, 1777; m. Sallie Davis; d. Morristown, N. Y., Feb., 1842.

Sylvanus (7), b. May 29, 1780; m. Rhoda Hathaway; d. Groton, N. Y., August 22, 1853.

Jonah Carpenter, b. Oct., 1744; m. Tervish Whitmore, in Ashford, Conn. Their children were:

Asa (7), b. October 10, 1770; m. Erepha Grow. She d. Dec., 1842.

Joseph T., b. January 2, 1774; m. Huldah Davidson, April 15, 1800. He died April 11, 1805, in Ashford, Conn.

Jonah, b. January, 2, 1774; m. Hannah Rice, Waterford, Vt.

Chester, b. July 3, 1780; m. Chloe Holt, March 16, 1815. She d. October 24, 1819, in Willington, Conn.

Isaiah, b. June 29, 1783; m. Caroline Bugbee, April 21, 1808, in Waterford.

Dyer, b. April 22, 1786; m. Martha Gibbs, September 19, 1811.

Alatheia, b. September 19, 1772; m. Abial Cheney, May 11, 1797. He d. September 16, 1841, in Waterford, Vt.

Dr. Cyril Carpenter (6), son of Comfort and Huldah, b. April 4, 1736; m., about 1759, Freelove Smith; d. December 9, 1816, a. 81. Freelove d. 1813, a. 76. Their children were:

Comfort A. (7), b 1760; physician at Pawtucket, R. L.; father of Genl Thomas Carpenter (8), of Providence, who was some years Democratic candidate for Governor of Rhode Island; Cynthia A., Cyril L., Benjamin B., Phebe T., Huldah H., b. Nov. 26, 1768, Christopher S., Polly S., Thomas O. H., b. 1777; Betsey M.

Orinda Carpenter (6), dau. of Comfort and Huldah, b. March 13, 1738; m., April 19. 1759, Nathan Dresser. Their children were:

Elfreda, b. October 16, 1759; m. Nathaniel Carpenter.

Huldah, b. October 18, 1761; m. (Abel ?) Jackson. She d. October 4, 1820.

Serena, b. February 26, 1764; m. Thomas Holmes; 2—Smith, of Ashford, February 13, 1841.

Esther, b. April 1, 1766; d. young.

Thomas, b. August 18, 1767; d. September 11, 1788.

Nathan, b. August 11, 1769; m. Rebecca Leffingwell. He d. May 13, 1834.

Mary, b. February 7, 1772; m. Ephraim Hyde.

Abel, b. January 26, 1775; m. Sally Brown.

Comfort, b. May 4, 1777; m. 1802 Celia Wade, born 1782.

Sally B., b. February 21, 1779; m. —— Sheller, of ——, N. Y.

Jonathan, b. January 8, 1782; m. ——; d. October 30, 1826.

Nathaniel Carpenter (6), second son of Samuel (5) and Hannah, b. November 20, 1718. His children were:

Amasa (7), Mary, Elijah, Nathaniel, Abishai, Eunice, Lucy, Hannah.

SEVENTH GENERATION.

Nathaniel Carpenter (7), son of Nathaniel (6) and Mary, b. July 18, 1756; m. June 20, 1775, Elfreda Dresser. He d. October 15, 1829. She d. May 11, 1840. Their children were:

Orinda (8), b. September, 1776; m. Thomas L. Haskell.

Martha, b. ——; d. young.

Abigail, b. August 15, 1781; m. Cap. C. Chandler. She d. April 16, 1849.

Harvey, b. March 12, 1784; m. A. S. Olmstead. He d. about 1852.

John, b. ——; m. ——, 1 —— 2, Nancy Well.

Thomas Dresser, b.——; m., about 1840, Lydia ——.

Abba (a son), b. ——; d. young.

Isaiah, m. Caroline Bugbee, April 21, 1808, in Waterford. Their children were:

Alonzo, b. April 22, 1809; d. November 18, 1809.

Caroline D., b. March 23, 1811; m. William Holt, March 21, 1836, Willington, Ct.

Sally B., b. February 13, 1813; d. February 13, 1813.

Isaiah P., b. January 22, 1814; d. August 10, 1840.

Sally M., b. May 19, 1816; m. Elijah, October 24, 1837, in Willington, Ct.

Amos B., b. May 25, 1818; m. Cosbi B. Parker, June 24th, 1847, in Lower Waterford, Vt.

Alatheia, b. January 11, 1821; d. July 18, 1821.

Ocena M., b. August 9, 1824; d. February 19, 1825.

Eliza, b. April 16, 1826; m. Jonathan Ross, November, 1852.

Alonzo P., b. January 28, 1829.

Abel Carpenter, of Rehoboth, Mass.; m. Olive Bliss (who was born May 29, 1763), Nov. 9, 1786.

Daniel Carpenter, of Attleboro', Mass.; m. Hannah Bliss (who was b. June, 19, 1774), Feb. 1, 1795, and removed to Genesee, N. Y.

Abel Carpenter, of Rehoboth, Mass.; m. Abby Williams Bliss (b. March 19, 1805), June 13, 1826. He d. Oct. 19, 1852. Children were:

George Nelson, b. March 25, 1827; d. May, 1827.

George Hodges, b. June 6, 1828; m. ——

Sarah Allen, b. Nov. 1, 1833; m. ——

Rachel, b. Nov. 28, 1835; d. March 10, 1852.

George H. Carpenter (son of Abel); m. May 20, 1860, Elizabeth P. Hunt. He d. without issue March 10, 1867.

Joseph R. Carpenter, of Rehoboth, Mass.; m. Sarah Allen Carpenter, daughter of Abel Carpenter, May 19, 1864. Had one child.

Frederic Howard, b. May 10, 1865.

Abraham Carpenter, of Rehoboth, Mass.; m. Elizabeth Bliss, Feb. 28, 1759. Removed to Vermont.

Lewis Carpenter; m. Mary Ann Bliss, Aug. 23, 1789, of Rehoboth, Mass.

James Carpenter; m. Dorothy Bliss, June 26, 1690.

Harrison Carpenter, of Savoy, Mass.; m. Harriet A. Bliss (who was b. May 22, 1829), May 25, 1851. Their child was:

Clara M., b. Aug. 20, 1854; m. Nov. 26, 1874, John M. Crofts.

Dr. Harvey Lessiams Carpenter (son of Dr. Nelson Carpenter and Eliza Lessiams); m. Mary Louisa Bliss (b. Oct. 14, 1833), Oct. 30, 1856. He was b. April 4, 1829; d. in Worcester, Mass., Feb. 2, 1875. Their children were:

Frank Nelson, b. Jan. 12, 1858; Herbert Bliss, b. Nov. 9, 1860; Walter Lessiams, b. Jan, 13. 1864; d. Sept. 14, 1866; Mary Gage, b. Nov. 10, 1868; d. Oct. 3, 1877.

Henry L. Carpenter, son of Wheaton and Alice Carpenter, b. in Attleboro, Mass.; m. Chloe M. Bliss (b. June 12, 1844), Nov. 20, 1868. She d. Aug. 5, 1870, and he m. a —— Waldron.

John L. Carpenter, son of Lewis and Mary C., of Fall River, Mass.; m. Sarah Durfee Bliss (b. Aug. 20, 1851), Feb. 2, 1873. Child was: Arthur Lewis, b. Jan. 23, 1874.

Merlyn Carpenter; m. Fidelia K. Bliss, March 22, 1854. Had 4 children.

Samuel S. Carpenter and Asena Bliss (b. Oct. 22, 1800); were m. July 25, 1830, (East Attleboro). No issue.

James Carpenter, of Rehoboth, Mass.; m. Lucy Bliss (who was b. June 23, 1769), March 26, 1788. He d. Oct. 20, 1812. She d. Sept. 21, 1817. Their children were:

Joseph, b. Sept. 8, 1789; Sarah Martin, b. Aug. 22, 1791; Lucy, b. May 23, 1794; m. John Mason; Rebecca, b. July 7, 1796; d. Sept. 19, 1810; Rosella, b. Aug. 26, 1799; d. Oct. 3, 1806; James, b. June 12, 1802; Newton, b. July 27, 1805; d. 1877-8.

Joseph Carpenter, son of the above James and Lucy: m. Feb. 21, 1813, Nancy Mason Bullock, who was b. Dec. 10, 1793, and who d. May 4, 1880. Their children were:

James Mason, b. Nov. 11, 1813; m. ——

George Moulton, b. Aug. 6, 1815; m. ——

Nancy Mason, b. June 14, 1818; m. Frances W. Garlin; Sarah Martin, b. Feb. 21, 1820; unmarried.

Jonathan Bliss, b. April 25, 1822; m. ——

Lucy Bliss, b. Aug. 1, 1824; m. Everett L. Sweet, —— 6, 1851; William Wallace, b. Nov. 8, 1826; Samuel, b. Feb. 26, 1829; m. ——

Jane Buffum, b. Feb. 26, 1829; d. Dec. 4, 1830.

Newton Francis, b. April 27, 1830; m. ——

Jane Buffum, b. May 23, 1834; d March 17, 1836.

Gouph, b. Dec. 22, 1835; d. Nov. 14, 1836.

Albert Norton, b. Aug. 14, 1837; d. Aug. 2, 1838.

Edward Everett, b. Oct. 2, 1840; m. Dec. 31, 1865, Emma B. Wilbur, and had one son, Jonathan E., who died young.

James Mason Carpenter, son of Joseph and Nancy Mason, Ins. Agt. and Farmer, Pittston, Me.; m. Aug. 18,1840, Martha Jane Reed Bodge, and had one child, ——, b. June 4, 1843; d. Oct. 26, 1851.

George Moulton Carpenter, son of Joseph and Nancy Mason, Ins. Agt., Providence, R. I.; m. July 10, 1843, Sarah Lewis Walcott, who d. March 6, 1869, leaving two children, George Moulton, Jr., b. April 22, 1841, a lawyer in Providence, R. I., and Edmund Janes, b. Oct. 16, 1845, editor at Central Falls, R. I.; m. Nov. 12, 1873, Lydia Etta Snow, and had 4 children.

Jonathan Bliss, son of Joseph and Nancy Mason; b. April 25, 1822;

m. Feb. 26, 1846, Lydia Ann Walcott. He d. Dec. 1, 1857, leaving one son, John Walcott, b. July 10, 1847; m. Dec. 10, 1873, Sarah B. Fuller, and had Joseph and Annie and two other children.

William Wallace Carpenter, son of Joseph and Nancy Mason; removed West; m. Jan. 1, 1854, Marinda Davis, and d. May 18, 1877, leaving issue; Sarah Martin, b. Oct. 9, 1854; Joseph William, Jane Buffum and Amy Jane.

Samuel Carpenter, b. Feb. 26, 1829, son of Joseph and Nancy Mason, a farmer in Cumberland, R. I., m. April 12, 1852, Ruth Ann Miller. Their children were:

Samuel Eber, b. Sept. 25, 1853.

Abby Laura, b. June 5, 1859.

Nancy Bishop, b. Feb. 21, 1864.

Newton Francis Carpenter, son of Joseph and Nancy Mason; a lawyer in Menomonee, Wis.: m. Dec. 7, 1851, Helen M. Brown. Children were:

William Francis, b. April 15, 1852; Nancy Mason, b. Jan. 2, 1858; Mary Elizabeth and Helen Maria.

The father afterwards married Esther Row and had three sons, George M., Edward Francis and Freddy Edson.

Richard Carpenter, of Amesbury, England. (Amesbury, in Wiltshire Co., England, is situated 7½ miles north of Salisbury, and 78 miles south-westerly from London. The town is of great antiquity, but has now but little trade. The Wesleyan Methodists have a meeting house here. Addison the poet was born here in 1672. Stonehenge is two miles west of the town). Richard was born 1575, and was buried September 21, 1625. Vide Register of Burials of Amesbury Parish, Salisbury, Wilts., England. John Selwin, minister.

William Carpenter, eldest son of the above Richard (and who was probably cousin to William, of Rehoboth), came to America and settled at Providence, R. I., with Roger Williams, 1636. (Swore allegiance in Providence, in 1666.) He married Elizabeth Arnold, a sister of Governor Benedict Arnold, the first Governor of Rhode Island. He was one of the principal men of the settlement, Member of the Council, Adviser, etc., and tradition says a preacher, as well as a founder of the first Baptist Society of Providence. He died Sept. 7, 1685. In his will, dated Feb. 10, 1679-80, he mentions "my eldest son Joseph and my daughter Lydia Smith, my daughter Priscilla Vincent (who, on May 31, 1670, married William Vincent), my two sons, Silas and Benjamin, my son Timothy, my son Ephraim, and the oldest son Ephraim of my son Ephraim, by his first wife," also my brother, Stephen Arnold, my grandson Anillion Carpenter, to Elizabeth, my beloved wife.

He also mentions Susanna, sister to his grandson Ephraim, who was not 21 years of age at the date of his will.

In the codicil, March 15, 1683-4, he revokes a part of his will to his son Ephraim, and wills it to his grandson Ephraim and his sister Susanna.

By a deed of gift to his sister Fridgsweet Vincent, he mentions his father Richard, who left property to him in Amesbury, in the County of Wilts., England, to wit: "a house in Frog Lane." Deed dated

December 4, 1671. In the will of Ephraim Carpenter, Jun'r, he mentions his father Ephraim, of Long Island. Ephraim, Jr., died Feb'y 22, 1697-8.

Joseph Carpenter, eldest son of the above William, was born probably in England. He remained at Providence with his father till about 1664, when he removed to Long Island, and on May 24, 1668, he purchased from the Indians the land known as Mascheto Cove, and the sale was confirmed unto him by Gov'r Edmond Andross, Sept. 29, 1677. The Oyster Bay records contain a great number of deeds to and from him, as well as mention of him as referee in disputes, &c. He married Ann Wickes. He died in the summer of 1683. Letters of administration were granted to his widow Ann and his son Joseph, July 9, 1684. His children were: Joseph, Nathaniel, William, Thomas, Benjamin and John.

Silas (2), brother of the above, married Sarah Arnold, daughter of Stephen Arnold. Silas died in Providence, Dec'r 25, 1695 (Sw. Alleg., 1671. His widow married, for her second husband, Edward Potter). The children of Silas were: Silas, William, Mary, Ephraim, Joseph, Israel and Jacob.

Joseph (2), son of Joseph (1), married Anne ——, and died about 1692, leaving sons Joseph (3), born Oct. 16, 1685, and Thomas.

William, son of Joseph (1), married Elizabeth ——. Their children were: William, Benjamin, Silas (and probably a Joseph), and one daughter Sarah, who married John Cook.

Nathaniel Carpenter, son of Joseph (1), married Tamar Coles. Their children I cannot fully trace, but am satisfied that Robert, "the Miller," Timothy, "of North Castle," Benjamin, "the Saddler," and John, "of Fredericksburgh," and Nathaniel, Jr., were his. He removed to North Castle, Westchester County, in 1719-20, and died there Feb'y 25th, 1730. (Record says Nathaniel, Jr.).

Benjamin, son of Joseph (1), married Mercy Coles (a sister of Tamar). Their children were: Joseph, of Lattingtown, b. Sept. 15, 1705; Benjamin, of Orange County, b. Nov. 3, 1708; Samuel, John and Timothy—five brothers—who all settled in Orange and Ulster County, N. Y., and two daughters (twins), Elizabeth and Hannah, b. Aug. 17, 1708.

John, son of Joseph (1), married Martha Feakes, June 12, 1713; they resided at Red Springs, L. I. (as did also his brother Benjamin). The children of John were probably John, Jr., Jacob, Joseph, Isaac, Martha and Phebe.

Benjamin, of Providence, son of William (1) (Sw. Alleg., May, 1671). Perhaps he married Renew, daughter of William Weeks, of Dorchester, but he was a permanent resident at Rehoboth. Their children were: Jotham, b. June 1, 1682, bapt. July 1, 1683; John, bapt. June 21, 1691, and Submit, Nov'r 5, 1693. All, I judge, at Dorchester, in right of their mother.

Timothy, son of William (1), married ——, and had issue, viz: Timothy, Elkalannah, Elizabeth, Hannah.

Ephraim, son of William (1), married ——, and had issue, viz: Ephraim, Susannah, Josias, and Joseph and Phebe.

THE CARPENTER FAMILY OF LONG ISLAND AND NEW YORK.

About the year 1540, one Cotleb Zimmerman emigrated from Prussia to England, where he married and settled. The name changed into English is Caleb Carpenter. He had a family of children, one of whom was named Ezra, who also married and had issue; among others, Elihu, who married young, and reared a large family. When he was quite advanced in years, the Society of Friends or Quakers arose, with whom he joined, and in consequence he was greatly persecuted and maltreated. He finally fled with his family, children, and grandchildren, taking refuge in Holland, where he remained a considerable time, until the persecution abated, when most of them returned to England.

From the above circumstances, it was supposed that the Carpenter family originated in Holland, and it is believed that some of them remained in Holland, and they, with other English exiles, established Quakerism there.

Ezra Carpenter, of Wales, England, born in the year 1570, had two sons, as follows:

1. Richard, b. May 4, 1593; d. June 11, 1669.
2. William, b. August 23, 1601; d. a bachelor in London, 1700.

Richard Carpenter, son of Ezra, m. Rachel ——, who was b. Feb'ry 27, 1601. Their children were:

1. Ephraim, b. Nov. 28, 1623.
2. John, b. Sept. 29, 1627.

John Carpenter, son of Richard and Rachel, m. Lois Hope. Both d. leaving no issue.

Ephraim Carpenter, son of Richard and Rachel, m. Elma ——, of Wales, England, who was b. June 17, 1627. Their children were:

1. Ephraim, b. Nov. 3, 1653.
2. Phœbe, b. July 24, 1658; left no issue.
3. Josias, b. Sept. 12, 1661; left no issue.
4. Timothy, b. Dec. 19, 1665.

[This family of four emigrated to America in April, 1678. Phœbe and Josias returned, and died leaving no issue.]

Timothy Carpenter, m. Mercy Coles, of Glen Cove, Long Island, N. Y., who was b. Feb. 2, 1668. Their children were:

1. John C., b. June 13, 1690.
2. Huldah, b. Dec. 18, 1692.
3. Jeptha, b. Dec. 29, 1693.
4. Benjamin, b. March 25, 1696.
5. Timothy (2), b. Jan. 4, 1698.

Timothy Carpenter (2), m. Phœbe Carpenter (who descended from the Rhode Island Carpenters), who was b. March 16, 1700. Their children were:

1. Samuel, b. Jan. 1, 1720.

2. Ephraim, b. July 27, 1723.
3. George, b. Aug. 7, 1726.
4. Phœbe, b. Jan'y 21, 1729; died without issue.
5. William, b. April 5, 1731.
—6. Archeleus,* b. April 23, 1734.
7. Silas } (Twins), b. July 15, 1737.
8. Benjamin }   b. July 15, 1737.
9. Timothy (3), b. Aug. 1, 1740.
10. Elizabeth, b. Nov'r 10, 1743.

William Carpenter, son of Timothy (2), m. Sarah Seaman, of Glen Cove, Long Island, N. Y., who was b. Nov. 7, 1735. William d. June 6, 1814. Sarah, his wife, d. Jan'y 1, 1791. Their children were:
1. Seaman, b. Feb'y 7, 1760.
2. Zeno, b. May 8, 1762.
3. Stephen, b. April 29, 1764.
4. Elizabeth, b. Sept'r 17, 1766; m. Southwick.
5. Bethana, b. Dec'r 5, 1767; m. Warden; died without issue.
6. Phœbe, b. March 23, 1769; m. Hoage.
7. Mary, b. Sept'r 23, 1771; m. J. Connell.
8. Caroline, b. March 8, 1773; d. without issue.
9. James, b. July 14, 1775; d. young.
10. Sarah, b. April 6, 1777; m. D. Carman.

Seaman Carpenter, son of William, m. March 19, 1791, Sarah Simmons, of Saratoga County, N. Y., who was b. August 30, 1771. Sarah d. Sept. 19, 1806. Seaman d. Jan'y 22, 1842. Their children were:
1. John, b. Dec'r 21, 1793.
2. Sarah, b. Jan'y 20, 1797.
3. Ruth, b. Oct. 14, 1799.
4. Hiram, b. Dec'r 14, 1801.

Ruth Carpenter, daughter of Seaman, m. Asa Barker, of Barkersville, on Nov. 27, 1817, who was b. Oct. 7, 1794. Their children were:
1. William C., b. March 14, 1819.
2. Susan M., b. Oct. 30, 1821.
3. Lydia Ann, b. Jan. 25, 1822.
4. Sarah, b. Jan. 30, 1825; died.
5. David, b. Sept. 17, 1827.
6. Mariat, b. April 25, 1831; died Aug. 19, 1878.

Asa, d. April 21, 1868; Ruth, his wife, d. Aug. 22, 1867.

Benjamin † Carpenter, son of Timothy (1), m. October 30, 1718, Dianah Alvenson, who was b. March 19, 1698. Benjamin d. March 26, 1778. Dianah d. Nov. 2, 1758, and was buried at Friends' Meeting House, in Chapaqua, Westchester County, N. Y. Their children were:
1. Eliza, b. Sept. 12, 1719.
2. Elijah, b. Dec. 23, 1722.
3. Ezra, b. May 6, 1726.

---

* On account of his loyalty, he emigrated to Nova Scotia during the Revolutionary War, where the family have since resided.
† Benjamin married a second wife, Lydia ——, who died Nov. 25, 1778.

4. Luther, b. Aug. 16, 1730.
5. Sarah, b. July 11, 1734.
6. Caleb, b. Sept. 25, 1736.
Caleb Carpenter, son of Benjamin, m. July 22, 1759, Amy ——.
Caleb d. Dec. 20, 1826. Amy d. Jan. 18, 1795. Their children
were:
1. Lebe, b. July 4, 1760.
2. Benjamin, b. April 1, 1762.
3. Mary, b. May 26, 1767.
4. Lydia, b. Aug. 4, 1769; d. 1796.
5. John, b. Oct. 20, 1771.
6. Zeno, b. Dec. 8, 1773; d. 1795.
7. Ruth, b. Jan. 24, 1776.
8. Caleb, b. Oct. 24, 1778; d. 1814.
Elijah, son of Benjamin, married Ellen ——, who was b. June, 18,
1728. Their children were:
1. Samuel, b. Oct. 5, 1751.
2. Amy, b. Jan. 3, 1753.
3. George, b. July 17, 1754.
4. Benjamin, b. June 30, 1756.
5. Phœbe, b. Dec. 22, 1760.
John Carpenter, son of Timothy 1st; married Cynthia ——, who
was born March 22, 1693, and died May 24, 1776. John died March
19th, 1771. Their children were:
1. John, b. July 1, 1714.
2. Lucretia, b. April 6, 1719.
3. Abel, b. Dec. 4, 1726.
4. Susan Ann, b. Sept. 17, 1730.
John Carpenter, son of John; married ——. Their children were
1. Daniel, b. Nov. 20, 1730.
2. Abraham, b. Dec. 27, 1738.
3. Nancy, b. Nov. 2, 1740.
4. Jesse, b. Dec. 18, 1743.
5. Jacob, b. March 6, 1745.
6. Zeppy, b. April 16, 1749.
7. Isaac, b. Sept. 3, 1751.
8. Gilbert, b. Sept. 4, 1754.
9. Sarah, b. Oct. 23, 1755.
Abraham Carpenter, son of John, the hatter; married Lydia Tot-
ten. Their children were:
1. John, b. Dec. 12, 1761; d. 1762.
2. Stephen, b. March 5, 1763; d. 1843.
3. Peter, b. March 15, 1765; d. 1833.
4. Anna, b. Dec. 12, 1769; d. 1770.
5. Freelove, b. Oct. 12, 1767; d. 1835.
6. James, b, Oct. 4, 1771; d. 1858.
7. Elizabeth, b. Jan. 22, 1773; d. 1848.
8. Daniel, b. May 30, 1775; d. 1840.
9. Abraham, b. Sept. 10, 1777; d. 1838.
10. Isaac, b. Oct. 10, 1779; d. 1836.
11. Jacob, b. Oct. 10, 1779; d. 1832.

12. Thomas T., b. Jan. 3, 1782; d. 1836.

Daniel Carpenter, son of Abraham: married **Fanny Hawkshurst** in 1799. Their children were:
1. Edward, b. Sept. 28, 1800.
2. Asa B., b. Feb. 11, 1802.
3. Daniel H., b. Nov. 2, 1806.
4. Phœbe Jane, b. Nov. 27, 1813.

Phœbe Jane Carpenter, daughter of Daniel, married, March 24, 1840, James H. Mills. Their children were:
1. William G., b. Sept. 1, 1841.
2. Frank H., b. June 7, 1844.
3. Charles C., b. Nov. 8, 1851.

Ephraim Carpenter, brother of Timothy (1), who was born in Wales, England, and who came to America in 1678; had a son, Ephraim Carpenter, who married, Nov. 15, 1675, Phœbe Hope, who was born Feb. 14, 1655. Their children were:
1. Amy, b. Oct. 8, 1676.
2. Josias, b. Feb. 10, 1681.
3. Joseph, b. May 29, 1684.
4. Julia Ann, b. June 21, 1687.
5. Ashman, b. Aug. 11, 1689.
6. Hope, b. Dec. 12, 1690.

Josias Carpenter, son of Ephraim, m. ————. Their children were:
1. Silas, b. Dec. 20, 1709.
2. Reuben, b. April 18, 1713.
3. Samuel, b. ——, 1716.
4. Elmara, b. Aug. 13, 1719.
5. Oliver, b. March 17, 1722.
6. Lucy, b. June 22, 1726.

Samuel Carpenter, son of Josias, born on Long Island, married, 1737, Elizabeth Leeds, of Egg Harbor. He died Aug. 2, 1804. Their children were:
1. Elizabeth, b. April 2, 1733.
2. Hannah, b. March 9, 1740; died.
3. Anne, b. Nov. 8, 1741; died.
4. Anne, b. Aug. 27, 1743.
5. Joseph, b. Aug. 14, 1745.
6. John, b. Oct. 14, 1747.
7. Hannah, b. Feb. 14, 1750.
8. Sarah, b. July 9, 1752.
9. Ruth, b. March 11, 1755.
10. Mary, b. Feb. 3, 1758.
11. Rachel, }
12. Sarah, } b. June 29, 1761.

Joseph Carpenter, son of Samuel, married ————. Their children were:
1. Joseph, b. June 15, 1769; died.
2. Sarah, b. Oct. 15, 1770; died.
3. John, b. July 17, 1772; died.
4. Susannah, b. April 8, 1774; died.

5. Samuel, b. March 2, 1776; died.
6. Jonathan, b. Jan. 6, 1780; died.
7. Ruth, b. April 24, 1783.
8. Rachel, b. July 25, 1786.
9. Elizabeth, b. April 10, 1790.

Samuel Carpenter, son of Joseph, married ————. Their children were:

1. Reuben, b. Jan. 13, 1802; d. June 14, 1802.
2. Joseph, b. Oct. 19, 1803.
3. Esther, b. Sept. 24, 1813.
4. Charles, b. Sept. 29, 1815: died.

Oliver Carpenter, son of Josias, married Martha ————, who was born June 8, 1724. Their children were:

1. Lydia, b. Sept. 3, 1744.
2. Barlow, b. Sept. 11, 1747.
3. Oliver, b. June 21, 1754.

Ashman Carpenter, son of Ephraim, married Lucy Amelia ————, who was born March 19, 1691. Their children were:

1. Silas, b. April 8, 1713.
2. Benedict, b. Jan. 11, 1715.
3. Archibald, b. May 16, 1717.
4. Margaret, b. Sept. 16, 1720.

Archibald, son of Ashman, married Hannah ————, who was born Dec. 26, 1721. Their children were:

1. Ashman, b. Aug. 27, 1741.

Ephraim and Josias acquired title to lands.

According to Thompson's History of Long Island, Ephraim and Josias purchased land in the town of Oyster Bay, on Long Island, N. Y., on the 9th of January, 1685, some six years after their arrival in this country. On the 26th of May 1663, the Indians sold a part of Mantinecock to Capt. John Underhill, John Frost and William Frost; another part on the 20th of April, 1669, to Richard Lating; another part, on the 1st December, 1683, to Thomas Townsend, and upon the 9th of January, 1685, the chiefs, namely, Sacanemen *alias* Runusuck Chechayen, *alias* Quaropin Samase (son of Tackapausha), being empowered by the rest of the Indians, conveyed the residue of Mantinecock, together with some other lands, for the price of sixty pounds of current merchantable pay, to James Cook, Joseph Dickenson, Robert Townsend, Stephen Birdsall, James Townsend, Daniel Weeks, Isaac Doughty, John Wood, Edmund Wright, Caleb Wright, John Wright, William Frost and John Newman: and thereupon the said grantees agreed to accept as joint purchasers with them the following named persons, being then the acknowledged inhabitants and freeholders of the town, comprising the most complete list of names at the time which the records present.

Among some forty-eight names appear the names of Josias and Ephraim Carpenter.

It appears that Ephraim and Josias Carpenter purchased lands in Oyster Bay, which joins Hempstead on the north.

Timothy Carpenter, when he arrived in America, was only thirteen years old, and was too young to purchase land or participate in municipal affairs. It appears he engaged in the cooperage business, in which his grandson, William, and his great-grandson, Seaman, also engaged in. Timothy Carpenter, being a Quaker or Friend, was not allowed to have any kind of monument erected to mark his place of burial. No family plots for burial were allowed; they were interred in rows, without regard to relationship, side by side. For this reason it is impossible to find his burial place. The town records of Hempstead, Long Island, were destroyed by fire on the 31st of October, 1797.

Timothy Carpenter, second son of the above Timothy 1st, removed from Hempstead, some fifty miles distant, to North Castle, in the county of Westchester, N. Y., where he purchased a farm, on which he resided until his death, and was interred in the Quaker burial-grounds at Chapaqua. His will was dated July 21, 1763, admitted to probate August 30, 1769, and is recorded in the Surrogate's office in the city of New York.

The above has been furnished by Mr. W. C. Barker. It is believed that it is a branch of the Massachusetts family, as Joseph, the eldest son of William, of Providence, settled on Long Island in 1664.

To the Members of the Carpenter Fund Association.

For years past it has been asserted and believed that there was in England a sum of money, variously estimated from two hundred to two hundred and fifty millions of dollars, which belonged to, and could be recovered by the lawful heirs of an ancestor named *William Carpenter.*

The origin of the family and the fund has been at all times differently stated, but the version given to me was that which had been published by former delegates, and other apparently reliable sources, which I quote as follows :

'The history of the Carpenter claim, now in the Bank of England, in brief, is as follows :

William Carpenter died a bachelor, at London, in the year 1700, at the age of 97 years, leaving an estate by will to his American heirs, valued at £40,000,000, or in U. S. Currency $200,000,000. Several efforts had been made to recover this immense fortune, and in the year 1845 an English branch of the Carpenter family made claim to this estate, the same being contested before the Lord Chancellor, at London, in the year 1846 ; occupying the court some five or six months.

This claim was made in the name of Henry Carpenter and Ashman Hope, of Manchester, England.

They did not claim to be heirs in the regular line of descent, but petitioned the Lord Chancellor to award the estate to them as a collateral branch, insomuch as the estate had been held in trust by the Crown for over one hundred years, and as the heirs to whom it rightfully belonged had not claimed it.

Their claim was not entertained. The following particulars were ascertained during the above proceedings :

On the 19th day of August, 1707, there was turned over to the custody of the Crown £2,796, 8, 11, which was invested in British Securities, at 3 per cent., and in addition to the above, it is said there still remains assets sufficient, when added to the above, to make a total of £40,000,000.

At the termination of the above suit a delegation of American gentlemen was empowered to contest the claim for some claimants who then resided in the vicinity of Boston, Mass.

They petitioned the Lord Chancellor to be granted a hearing, and to be allowed the use of the testimony produced by the Manchester Carpenters, so far as it might be applicable to their case.

The petition being granted, the proceedings were then continued in the interest of the Boston Carpenters, who, however, were unable to establish their line of descent.

Their suit was denied, as it was found, in the language of the Lord Chancellor's decree, "That the claimants were not the lawful and lineal descendants of William Carpenter, who died in 1700."

One of the said delegates stated that it was admitted that the rightful heirs to the estate of William Carpenter are the descendants of Ephriam, Josias, Timothy, and their sister, Phœbe Carpenter, who came to America in the year 1678.

In addition to the above, in an estate left by William Carpenter, of Providence, R. I, amounting to one hundred and sixty thousand pounds. Will dated 1684.

There are other Carpenter estates in the Court of Chancery unclaimed.

No details as yet have been ascertained, but it is known that they number six or more. The small estates were left between 1684 and 1801, and the several amounts at the time they were left were as follows:

£2,700; £1,600; £1,900; £2,300; £2,700; £3,100; £19,100.

It was stated that the different amounts above named belonged to the various branches of the Carpenter 'family in America. Other parties interested had other theories as to the funds, estates and ancestors; but as these were not made known to me, and as I had no other knowledge, I was compelled to adopt the above theories, as they came from apparently authentic sources. It was also asserted that advertisements had repeatedly appeared in the English newspapers calling upon the heirs of William Carpenter to make good their claim.

It was further said that the money had been invested in annuities, and that the annual list had been published by the English authorities invariably calling for the said heirs, and was further stated that lists of unclaimed money in the Bank of England and Court of Chancery had frequently been published, and that the name of William Carpenter always appeared in them.

So convincing has been the belief in the existence of this fund, that several associations composed of the descendants of Carpenters (your own among them) had been formed and sent delegates to England to make the requisite investigation and to institute proceedings necessary to recover it.

Among these Mr. Frederick Arsdale and Mr. William Carpenter, of Philadelphia, Pa., the foremost. They made elaborate statements which were the results of their investigations, which they transmitted to and were received by the different members of the Carpenter family as entirely reliable.

COPIES OF LETTERS FROM MR. PHILLIPPS, A CLAIM AGENT OF LONDON, ENG.

No. 93 HIGHGATE ROAD, N. W., LONDON, 17 July.

I have had the pleasure of receiving your letter of the 22 ultimo, inclosing an article from a newspaper giving a detailed history of the Wm. Carpenter claim. I am in a position to confirm the truth of the statement therein set forth. The money is all right and can no doubt be recovered, but in the first place you must send me all the particulars of your client's pedigree. The government alleged that Wm. Carpenter, who died in 1700, was an illegitimate, but this is not the case ; and if your client's pedigree is all right I think we can go in and win. But you had better send in the particulars at once, so that if there is any defect in the pedigree I may be able to make it good. The next thing to take into consideration is the money which will be required to fight the battle with. However, supposing your clients are poor, if the pedigree is correct I shall have no difficulty in raising the necessary funds. It will be best, however, before deciding upon the course to be pursued, for me to certify as to the correctness of the claim.

Yours faithfully,

(Signed,) JAMES PHILLIPPS.

In answer to the above I sent a large pedigree, and also requested that a statement should be forwarded to us, but in response received the following letter :

LONDON, Oct. 29th, '82.

If you require any further information from me you must send me £100 or $500, because, if I would send you the information you ask for, which is of great value, you would, no doubt, serve me as I have been often served by Americans.

You will also please observe that, if you expect me to write you on this case, you must send me £10 or $50 to cover time and postage. I have already devoted much time to this business.

(Signed,) JAMES PHILLIPPS.

In answer to this I stated that I simply wished to establish the fact of the existence of such an estate, known as the William Carpenter estate, unsettled and awaiting a rightful claimant, and if he had proofs which would do this beyond a reasonable doubt, I was prepared to pay him liberally for such facts. I stated that I did not want him to reveal any secrets to me or give me any of his facts till the proper time.

But, is there a case, and can you establish the fact beyond a reasonable doubt? I asked. In response, I received the following:

LONDON, Jan. 4th, 1883.

This is a very great estate and a great haul can be made, provided we have got the right parties, but all will depend upon the pedigree, which, upon the receipt of $500, I will make perfect, and have a proper legal opinion upon it before I open fire for the recovery of the property.

With respect to terms for the recovery of this vast sum of money, I think I ought to have half.

Yours truly, JAMES PHILLIPPS.

These reports seemed to be entirely accurate. In fact, there was no reason to doubt them. They were corroborated inferentially by all the accounts received from other sources, and the details of the investigations inspired confidence. Your association relied upon it, and no blame can be attached to you for such reliance. You would have apparently been false to the object and purpose of your organization, if you had not taken measures to protect what thus appeared to be the vital interests of your constituency.

Of course you had no means of judging of the correctness of the statements as to the genealogy, but the showing as to the fund was such as to demand action upon your part.

Believing that your interests should be represented by some one who was devoted to them exclusively, you honored me with an offer of employment to ascertain definitely whether such an unclaimed fund existed, and if it did, then to ascertain further to whom it belonged, and what would be the most expeditious and certain way of recovering it.

I accepted the offer, and after having exhausted such resources as I had in England, to no avail, I found it necessary to go to London, personally, so that I might give the case my personal and prompt attention.

The importance of this matter, and the great interest which it has for years excited, have impressed me with my responsibility and led me to examine the records with as much care as possible.

Although working under great disadvantages, I have not been deterred by fear or favor from walking steadily onward in the path of equity and justice.

I am convinced that an honest opinion and a true statement of the facts of the case, in all such cases, should in no manner be evaled, but should be fairly stated by the expression of one's careful, guarded and unbiased judgment; and if one has fairly and honestly acted, I believe he performs his duty and gives to his clients his best endeavors.

I will state such facts and information which I have been able to gather about the case, with frankness, candor, and sincerity, that those persons who know me would recognize as the controlling principle of my life, by calling things by their proper names.

I will carefully abstain from straining any point or over-estimating facts, even to a hair, for the purpose of showing the fallacy where I do not sincerely believe it to be.

I arrived in London after an unexpected delay, and began my task on the 18th day of April, 1883. The work was by no means speedily accomplished, as there were numerous obstacles and embarrassments in the way. Among the most serious of these were the official regulations of the various public offices, which amount almost to a prohibition of all inquiry. I quote them, viz.:

The rules prescribed by the Master of the Rolls respecting the Chancery Master's Documents in his charge and superintendence, passed pursuant to Stat. 40 and 41, Vic., Chap. 55.

I. "Any person wishing to inspect, or have copies of the Chancery Master's Documents, shall address a petition to the Master of the Rolls, stating the nature and object of the search."

II. "Except under special circumstances, no person will be allowed to inspect or have copies of the documents unless he satisfies the Master of the Rolls, by *prima facie* evidence, either that he claims, under one of the parties to the cause, who, if living, would be entitled to inspect the documents, or that he would be a person entitled to revive the cause, if it were effective, or that he has a right to the documents themselves.

III. "The Master of the Rolls, should he see no objections to the application, will then, subject to the consent of the Treasury being obtained by the parties, give the necessary permission.

IV. "The fees are the same as that of the regulations of the Master of the Rolls.

"24th May, 1875.   (Signed,)   G. Jessel, M. R."

This is followed by an application to the Treasury for its consent. Under this rule I was peremptorily refused permission to make a personal inspection of the books at the office of the Accountant-General in Chancery.

In the second supplement to the London Gazette, March 1st, 1877, is a like rule as to the funds now in chancery.

"No information is to be given by the Chancery Paymaster respecting the money or securities to the credit of any cause, matter or account in this list, until he has been furnished with a statement in writing by a solicitor requiring such information, of the name of the person in whose behalf he applies, and that in such solicitor's opinion the applicant is beneficially interested in such moneys or securities."

No information of any sort is volunteered in any of the public offices in England.

It is only to be got upon the most pointed and direct interrogatory, persistently repeated, and the response is as pointed and brief as possible.

All the very ancient documents are in either *Latin, Norman, French,* or *Black letter,* and often such a mixture as to be quite incomprehensible at first to one unused to reading them, and most of them are so obscure by age as to require much time and patience to decipher them.

My search, as will be seen, was largely among these folios. The subject would run together, and for days at a time I would be driven back and forth from one topic to another in the hope of obtaining the particulars demanded before a proper answer would be given to my inquiry.

For instance: In asking at the chancery office whether any funds existed which belonged to William Carpenter's heirs, it was required to be stated, before replying, what particular persons of that name claimed to be heirs, what relation they were, how and in what way I was interested.

This would drive me back to the genealogical researches, and thus I was continually retarded and embarrassed.

These were not the only discouragements. It was not long before I found that there existed a general distrust of such claims, and the Carpenter claim in particular was well known had been frequently examined by responsible lawyers, who had all agreed that there was nothing in it.

Messrs. J. & W. Maud, prominent English solicitors, to whom I am under many obligations for professional courtesy extended to me, had previously examined the claim thoroughly, and was entirely satisfied there was no fund, and that all my time given to the case would be uselessly wasted. Messrs. Cowlard & Cowlard, who were previously instructed by me to probe the matter to the bottom, sparing neither time or expense.

They made the search as directed, and reported that, positively, there is no such fund in existence to be paid, even if any heirs of William Carpenter could prove a right to receive it.

Messrs. Poole, Hughes & Poole stated that, after a thorough examination of the records, they were unable to find any record relating to the matter.

Messrs. Bumpas, Bishoff & Dobson, prominent English solicitors, informed me that they had made careful search at the Public Record office for any proceedings in the name of either Carpenter or Hope, for the years 1843 to 1848 inclusively, but could find no orders in any causes or matter in either of the names between those years in any way relating to the subject matter.

Other equally eminent solicitors—one an American, who was settled in London—considered the pursuit hopeless. He stated that the practice of advertising for missing heirs to appear and receive their immense sums of unclaimed money was a regular business in London, with branches established in most of the American cities, and regretted that his countrymen should be constantly defrauded by being made the dupes of such devices. He stated that he knew that there was no such fund, and never had been no such ancestor, no such claim, and no use in looking it up.

This, it was presumed, should be conclusive, and I might accept it at once, but I stated that I proposed to make a personal examination.

All these parties were of the highest integrity and professional standing, with no motive whatever to mislead, but, on the contrary, every desire to aid me.

If I had not pledged myself to make a personal examination, I would have accepted their statement and immediately returned, but I felt that I must go on, and would have been glad to have received from them any affirmative suggestions as to the early Carpenter genealogy. They could give me none. They could not give me any actual starting point in the genealogy. They had had none for themselves, and hence could not give me any data by which I could review their work or be ,in my own. All they knew was, that there was no such fund. After these interviews I ceased to ask for any further suggestions from any one. I saw that I must depend on my own exertions.

The whole field was untrodden. I found it a labyrinth which I was compelled to explore in darkness and ignorance, without a compass, and with obstructions and confusions on every side.

I did explore it, nevertheless, and now that I have done so, I am reluctantly and regretfully forced to admit that all they told me was entirely correct, and that my search was unprofitable, although not useless.

After months of conscientious and assiduous labor under the most trying circumstances, I have thoroughly satisfied myself that there is no fund now in the Bank of England or the Court of Chancery in the name of Carpenter, or belonging by descent to any member of this or other similar associations.

The motive which prompted this deception upon a large body of expectant people, wholly at that time without the means of detecting it, cannot be too strongly characterized. I also found many other of the statements made and published, to be equally unwarranted, and nearly all of them to be more or less misleading.

I next turned my attention to the calls for next of kin, and found this basis in them, in a book published by a Mr Preston, entitled, "Index to Heirs at Law, Next of Kin, Owners of Unclaimed Money, Missing Friends and Legatees or their representatives, in chancery suits, who have been advertised for during the last 150 years," &c., &c.

There is a numerical reference to twelve advertisements for persons of the name of *Carpenter*. I have procured a copy of this book, which is at my office for inspection. I procured the full facts as to each of these advertisements and give them herewith.

Carpenter, Daniel, late of Herningsham, in the County of Wiltshire, who died June 30, 1835; is supposed to have left surviving him a brother James and a nephew Thomas. Next of kin inquired for 21st Dec., 1854. This consisted of a small personal estate, which was divided amongst them in due course.

Carpenter, Ann, spinster, late of Eastbourne, in the County of Sussex, and who died April 30, 1857. Next of kin inquired for in 1858.

Carpenter, John and George, late of Shepherds Bush, Hammersmith, in the county of Middlesex; creditors inquired for in 1837.

Carpenter, James, late of Leigh, in the parish of Westbury, in the county of Wiltshire, and who died Dec., 1817. Inquired by Court of Chancery 1822.

Carpenter, George, late a General in Her Majesty's East India service. Inquired for by Court of Chancery 1855.

Carpenter, Henry, who died in London in 1837. Next of kin inquired for.

· Carpenter, J. & Co., inquired for to take dividends arising from the estate of Le Messurier Haviland, 1834.

Carpenter, John, late of Newton Lane, London. Deceased inquired for by the Court of Chancery in 1772.

Carpenter, John, who died at the Falkland Island. It is understood that he had relatives residing at Blackheath or Greenwich. Next of kin inquired for in 1847. The above amounted to £9.0.5 sterling, which sum was duly paid over to Sarah Webster of 4 Pollin St., London, who established her claim to the same.

Carpenter, Richard, of Newcastle Place, Edgewater Rood, in the County of Midd., coach builder. Creditors inquired for in 1854.

Carpenter, John, a butler in 1836 for Major Thomas Cope, of London. Next of kin inquired for 1847.

Carpenter, children of Dr. Nathaniel, late of the county of Kings and Queens, Virginia, who died in the month of April, 1778, leaving Corydon Carpenter, William Fountleroy Carpenter, Nathaniel and Bushrod Carpenter, him surviving or either of the next of kin or legatees of the said Corydon, William Fountleroy, Nathaniel and Bushrod Carpenter.

It was evident at a glance, that neither of the above related to the object of my search; but by the time I had discovered this, I had also discovered that it is a common custom in England to advertise in this manner for debtors and witnesses who have absconded or whose residences are unknown, and to endeavor to get service of summons in suit upon secreted defendants, and for a variety of similar purposes. Not one in five hundred is a genuine advertisement for the purpose professed.

Not long since there was an exposure of this swindling practice, which, for audacity and extent, even surpassed the frauds of the notorious adventuress, *Miss Furneaux*. A company styling themselves "The International Law Agency," avowedly established for the purpose of discovering heirs, next of kin, and recovering vast sums at present in chancery, announcing that they had important agencies in New York. Their advertisements, which were extensively circulated through every possible medium, prominently set forth that next of kin and heirs were wanted for unclaimed money amounting to £25,750,000, and offering to supply on receipt of a postage stamp, either a circular on unclaimed money, or an elaborately compiled book, containing names of persons wanted to succeed to unclaimed money and property.

Business was also invited by additional advertisements, requesting all "parties" bearing the commonest names, such for instance as Smith, Jones, Johnson, White, Brown, Davis, Wood, Ward, &c., to apply at the office.

These lists of names were regularly varied in the announcements, and the tempting bait seems to have been irresistible.

Many hundreds were victimized in this manner, although no public exposure of the deception practiced has, for obvious reasons, hitherto been made. Implicit confidence seems to have been shown by the army of dupes, with a few exceptions, and in these cases, the persons by threats and even personal violence succeeded in having returned to them the "fees" which they had paid to the agency upon the most glowing representations as to the untold wealth which rightly belonged to them, and which was easily obtainable by the agency. Inquiries have resulted in showing pretty clearly the methods of these swindlers.

In every instance, applicants for information were informed that they were entitled to property. A clerk engaged in the office stated confidentially that he had no knowledge of any money ever being obtained on behalf of clients through the instrumentality of the agency.

Although consultations were announced to be without charge, a guinea and a half or more appears to have been levied as preliminary expenses, after which the charges accumulated according to the circumstances, and upon the most ingenious pretexts.

Another fertile source of revenue to the agency, was the payment for a lithograph form of agreement, which each applicant was required to sign.

This agreement, which is skillfully drawn, pledges the client to pay to the agency a percentage—generally 20 per cent.—varying according to the amount of the claim.

Claimants were told time after time that the conductors of the agency believed their claims were incontestable, and that the agency knew where to put their fingers upon the money, and that they had been watching it a considerable time.

These statements were supported by the production of letters, purporting to contain official acknowledgments from the courts, of the identity of the claimants.

The entire proceedings were a sham, and not the slightest effort was made to carry out the professed intentions for which the offices purported to have been opened.

The wickedness of these hollow-hearted cheats and swindlers, who prey on credulous, ignorant and weak-minded people, are found in every part of the United States.

The grossest instance of humbug that I have met with for a long time, is that of an individual in this city, who advertises that he is an att rney at law, and the only office of record of unclaimed estates and money to be found in the United States.

This individual should be required to furnish information which would establish his right to the title of attorney at law, and how this ignoramous can take in intelligent people is a mystery to me.

There are several claim lawyers in New York City and elsewhere, who, under pretense of investigating claims, swindle those who confide in them. The method of investigation adopted by some of these so-called claim lawyers is outrageous. .

The hopes of the ignorant are fed, their pockets plundered, fees are charged and no service rendered.

In some such cases no service could possibly be performed, as the estates are entirely out of the sphere of action of said claim lawyers. Books are published by them containing lists of supposed heirs. These books are only bait to hook; those who "bite" will be "bitten," taken in and swindled.

A claim lawyer of the above description will always talk much about himself, tell you what great things he achieved, he recovered this and that.

Such men are very avaricious. and leave a bad impression, because of their meanness. They would take the last dollar from a poor widow for worthless service, and, in my opinion, they are as near thieves as the law allows. Age is no evidence of virtue or capability. One may be within a few years of his grave, and yet be selfish, mercenary, wicked.

No other rule can be given by which to judge such creatures, than may be found in their miserable faces and their bad heads—of course their advertisements, personal boastings. etc., will convince those having ordinary intelligence, of their worthlessness.

BANK OF ENGLAND.

This bank, the most important in the world, was projected by *William Patterson*, and was incorporated July 27, 1694. It was constituted as a joint stock association, with a capital of £1,200,000, which was lent at interest to the Government of William and Mary. at the time in a state of embarrassment, at its very outset; therefore, the Bank of England was a servant of Government, and, in a lesser or greater degree, it has enjoyed this character through all the stages of its subsequent history. At first the charter of the Bank was for eleven years only; but, in consequence of the great service of the iustitution to Government, its charter has been at various times renewed. Viewed in its banking department, the bank differs from other banks in having the management of the " Public Debt," and paying the dividends on it, in holding the deposits belonging to Government. and making advances to it when necessary, in aiding in the collection of the public revenue, and in being the *bank of other banks.*

The bank is also custodian of boxes deposited in its cellars for safe keeping. It is a pity that these boxes are not overhauled after a lapse of a certain number of years, and their contents advertised. It has occasionally *oozed out* that many of these consignments are not only of rare intrinsic and historic value, but of great romantic interest. For instance, some years ago the servants of the Bank of England discovered in its vaults a chest, which, on being moved, literally fell to pieces. On examining the contents, a quantity of massive plate, of the period of King Charles II., was discovered, along with a bundle of old love-letters written during the period of the Restoration.

The Directors of the bank caused a search to be made in their books. The representative of the original depositor of the box was discovered, and the plate and love letters handed over. There is also a large collection of articles deposited in the vaults belonging to suitors in chancery.

The following are a few of the most noticeable:

A box containing small articles of jewelry.

A box containing documents of titles, jewels, trinkets, watches and personal ornaments.

A box marked " diamond necklace, coronet and ear-rings."

A box containing plate and other articles.

A bag of clipped money, etc. (Jones v. Lloyd, 1726.)

A box marked "Securities for Legacies."

Two boxes containing plate belonging to a person of unsound mind.

The list comprises some eighty entries, doubtless of very great value in the aggregate.

The bank does not re-issue a note—when once paid it is canceled. £5 is the smallest note issued.

The stock of paid notes for five years is about 77,745,000 in number, and they fill 13,400 *boxes*, which, if placed side by side, would reach 2½ *miles;* if the notes were placed in a pile, they would reach to a height of 5⅜ *miles;* or, if joined end to end, would form a ribbon 12,455 *miles long.* Their superficial extent is rather less than that of Hyde Park, London. Their original value was over £1,750,-626,600, and their weight over 90⅔ *tons.*

Among things not generally known is the fact that there annually lapses to the English Government a very large sum from unclaimed dividends, presumably by reason of the representatives of the original stockholders not being known to the Bank of England authorities. I gather this information from a return presented to Parliament by the National Debt Commissioners during the session of 1877. Some readers may like to have the extract in full:

1876.		£	s.	d.
4th April.	Dividends due and not demanded........	984,225	4	7
	Advanced to Government..............	876,739	0	9
4th July.	Dividends due and not demanded.......	901,080	1	7
	Advanced to Goverument..............	876,739	0	9
4th Oct.	Dividends due and not demanded.......	898 516	4	3
	Advanced to Government..............	876,739	0	9
1877.				
4th Jan.	Dividends due and not demanded.......	885,508	4	11
	Advanced to Government..............	876,759	0	9

The sums thus advanced to the Government are applied, pursuant to the provisions of a certain Act of Parliament, toward the reduction of the National Debt. The unclaimed dividends accrue, doubtless, on a variety of stocks not all entitled to the same rate of interest. Most people reading the return from which the above figures are extracted would probably think that four separate sums of £876,-739 0s 9d. each were advanced to the Government during the year 1876-7. I confess I thought so myself on first looking at the return, but I have since been informed that the item "Dividends due and not demanded" includes all unclaimed dividends up to that date, and that only one sum of £876,739 0 9 has been advanced to the Government.

Having exhausted the advertisements, the next thing to be examined was the list of unclaimed dividends upon annuities, and upon which so much had been predicted.

In order to understand this list, it is necessary to state the law which governs unclaimed dividends upon annuities.

By an Act of Parliament, passed 56 George III., Chap. 60, it was provided that all capital stock upon which dividends should remain unclaimed for the space of at least ten years at the Bank of England . . . . should be by the bank transferred to the Commissioners for the Reduction of the National Debt, and the dividends upon it thereafter unclaimed should also be paid to them.

The first Act was afterward amended, 9th July, 1845, 9 Vict. I have the amended bills here for inspection.

Sec. I.—Provides, that after any such stock or dividends have been thus paid over, it may be repaid to claimants under certain conditions.

Sec. II.—Provides that three months' notice must be given by advertisement before transfer or payment of any stock or dividend to any claimant when the amount exceeds £20.

Sec. III.—Any person may apply to the Court of Chancery to rescind or vary any order before actual re-transfer or payment.

Sec. V.—The Lords of the Treasury may authorize inquiries into the circumstances of unclaimed stocks and dividends whenever necessary.

Sec. 52 provides that, immediately after every such transfer, the following particulars shall be entered in a list to be kept by the bank: (1) The name in which the stock stood immediately before the transfer; (2) The residence and description of the parties; (3) The amount transferred; and (4) The date of the transfer. Such list to be open for inspection at the usual hours of transfer; duplicates of each list to be kept at the office of the National Debt Commissioners.

Sec. 54 deals with subsequent dividends. It provides that where stock is transferred, all dividends accruing thereon after the transfer shall be paid to the National Debt Commissioners, and shall be from time to time invested by them in the purchase of other like stock, to be placed to their account of unclaimed dividends. All such dividends, and the stock arising from the public investment thereof, shall be held by those Commissioners for, and subject to, the claims of the parties entitled thereto.

Sec. 55 relates to re-transferred, and payment to persons showing title; it is in substance as follows:

Re-transfer may be made to any person showing his right thereto. In case the authorities are dissatisfied with the claimant's title, he may by petition, in a summary way, state and verify his claims to the Courts of Chancery, and the Courts may make such order thereon, touching the stock, dividends and costs of application, as to the Court seems just.

Sec. 59. It may be desirable to give this section *in extenso*, as it relates to cases where a second claimant appears :

Where any stock or dividends, having been re-transferred or paid, as aforesaid. to a claimant by either bank, is or are afterward claimed by another person, the bank and their officers shall not be responsible for the same to such other claimant, but he may have recourse against the person to whom the re-transfer or payment was made.

Sec. 60 provided that if in any case a new claimant establishes his title to any stock or dividends re-transferred or paid to a former claimant, and is unable to obtain transfer or payment thereof from the former claimant, the Court of Chancery shall, on application by petition by the new claimant, verified as the Court requires, order the National Debt Commissioners to transfer to him any such sum in stock, and to pay to him such sum in money or dividends as the Court thinks just.

Sec. 63 enables the treasury to empower the Bank of England or Ireland to investigate the circumstances of any stock or dividends remaining unclaimed, with a view to ascertain the owners thereof, and allow them such compensation for their trouble and expenses as to the Treasury seems just.

In 1870 all enactments relating to the National Debt were consolidated, but Part VIII. of the Consolidation Act re-enacts the above provisions.

Now as to any Carpenter annuities.

The first list was published in 1823. It states expressly that it includes " All funds and securities transferred . . . . on or before the year 1820.

Therefore it must have included any Carpenter fund, because it is beyond dispute that if any such fund had been in existence, it had certainly accumulated before 1820.

This list will be found in the Record of Transfers, published by order of the Directors of the Bank of England, A. D., 1823.

It is entitled:

The Names and Descriptions of the Proprietors of Unclaimed Dividends in Bank Stocks and all Government Funds and Securities transferred at the Bank of England, which became due on or before the year 1820, and remained unpaid the 31st December, 1822, including all stocks upon which dividends have been unclaimed for ten years together, which have been transferred and paid to the Commissioners for the reduction of the National Debt, pursuant to Act 56, Geo. III., Chap. 60, with the date when the first dividends respectively became payable, and number of dividends due.

There is no dividend or stock in the list which belongs to any Carpenter to whom you can trace, directly or indirectly.

The only Carpenters are these:

1719. Carpenter, Ann, St. Mary's, Whitechapel, spin.
1743. Carpenter, Mary, St. Dunstan West, deceased, 14 div., 4 per cent. annuities.
1744. Carpenter, Adrian, Bowser, Covent Garden, 3 div.
1745. Carpenter, Ester, St. Bennet's Fink, widow, 2 div.
1751. Carpenter, Henry, Stepney, 1 div.; old South Sea annuities.
1761. Carpenter, Thomas, London Bridge, 2 div.
1765. Carpenter, George Redburn Heris, E-q., 1 div.
1765. Carpenter, Joseph, Launceston, shopkeeper, 1 div.
1771. Carpenter, William (clerk), Launceston, 1 div.
1773. Carpenter, Joyce, Upper Toothing, widow, 1 div.
1774. Carpenter, Sarah, Stoke, Newington, spinster, 1 div.
1786. Carpenter, Thomas, Cateaton Str., warehouseman, 1 div.
1788. Carpenter, Eliz., Manchester Sq., spinster, 8 div.
1805. Carpenter, Philip, Chain Str., Westminster, dealer, 1 div.
1807. Carpenter, Robert, Bath, gent, 31 div.
1816. Carpenter, John, Maiden Lane, Covent Garden, gent, 1 div.
1820. Carpenter, Edward, Oxford, coachbuilder, 6 div.

As you had no interest in any of these people, it was, of course, useless to attempt to gain any further details as to them, especially under the stringent rules. The publication of this list settled definitely, in my mind, that there never had been by your William Carpenter, or any one for him, deposited any sum of money which had been invested in "British security," or any bonds whatever, which bonds had been transferred to the National Debt Commissioners.

If there was any such Carpenter fund at all, clearly it was not in the transferred annuities. There could be no dispute about that and there was no need of pursuing that branch of the investigation any further.

"CHANCERY OFFICE."

In the olden times the Masters in Chancery had the custody of all moneys and effects deposited in Court in suits referred to them, and the Usher took charge of any property brought into Court in suits which had not been referred to one of the Masters. The Masters and the Ushers were responsible for all moneys and other property received by them, and were bound to distribute the property so intrusted to them by orders of the Court. In the meantime they employed the money in their hands for their own benefit. This practice continued until the bursting of the South Sea Bubble, when it was found that several of the Masters were defaulters.

A remarkable trial (reported in State Trials, Vol. 16,) was the result, and ended in the then Lord Chancellor (Lord Macclesfield) being fined £30,000. The default of the Masters amounted to over £100,000 ; this default was made good by increased fees on the suitors, and stringent precautions were taken to prevent the recurrence of such a scandal. Each Master was directed by an order of the Lord Chancellor, dated 17th December, 1724, to procure and send to the Bank of England a chest with one lock, and hasps for two padlocks ; the key of the lock to be kept by the Master, the key of one of the padlocks by one of the six Clerks in Chancery, and the key of the other by the Governor or Cashier of the Bank. Each Master was ordered to deposit in his chest all moneys and securities in his hands belonging to the Suitors, and the chests were then to be locked up and left in the custody of the Bank, and to be so kept that the Masters might have easy access thereto under orders of the Court. This plan did not work well, for it was found that by the rules of the Bank of England the vault where the chests were kept could not be opened unless two of the Directors were present with their keys, and it soon was found that great trouble, difficulty and expense would be occasioned to the Suitors by requiring the attendance of no less than five officials whenever any of the chests had to be opened to deliver out effects and to receive the interest due. In 1725, therefore, a General Order was made directing that all money and effects should be taken from the Master's chest and given into the custody of the Bank. Duplicate accounts were to be kept at the Bank and at the Chancery Report Office, and any dealing with the. Suitor's money was to be certified to the Report Office.

Another General Order extended the plan to moneys and effects in the custody of the Usher of the Court.

In 1726, the first Accountant-General of the Court of Chancery was appointed, and all funds in the custody of the Masters or Ushers were transferred to his charge. An act of Parliament, passed in 1725, gives power to appoint an Accountant General, and contains elaborate provisions with reference to the custody and safety of the Suitor's Funds.

At the Chancery Office I applied to the proper department of the Accountant Generals and was informed that there was no William Carpenter fund held there in trust or otherwise.

The only ones mentioned are Coryndon Carpenter, who died July, 1776, and by his will, which was proved in the Court of the Archbishop of Canterbury on the 16th July, 1776, which gave an annuity of £50 to his brother Nathaniel Carpenter, and devised two farms at Webworthy, and North Treglan, to his executors in trust for sale.

The Testator's brother (Dr.) Nathaniel resided in Virginia, where he died in 1773, leaving four children him surviving, viz :—Coryndon, William Fountleroy, Nathaniel and Bushrod. After the death of the Testator's brother Nathaniel the property passed under the will to the said four children, in equal shares.

It appears that at the death of Dr. Nathaniel his four sons were very young and that the second and the third son attained their majority in the years 1796 and 1797.

In 1792, the surviving executors of the Testator's will exercised the power of sale as to Webworthy, and thereby realized £1,150, which of course belonged to the four children in fourths.

The eldest son, Coryndon, was then of age, and consequently received his fourth. The remaining three-fourths was handed to Trustees, and invested in consols.

This sum it was declared by a deed dated 4 Aug. 1792, was held by them as an indemnity to the purchaser against any claim put forth against him by the three remaining children, who were then infants. Their shares to be paid to such infants, upon their releasing any claim in respect of property so sold.

The eldest son joined in such deed, so as to release his claim.

The shares of the second and third son were paid to them on attaining 21, and thereupon they released their claims against the purchaser, consequently Bushrod, the youngest son, was the only one remaining, and he never applied for his share, and was last heard of in 1794. He was born in Virginia in 1783, and his share, by accumulations and new investment of the dividends, has increased from £309.5.11, in 1792 to £3.000.

William Fountleroy Carpenter, the second son, died in 1796, a bachelor, and administration was obtained upon his estate in 1797.

Bushrod is supposed to have died a bachelor about 1794.*

[Mr. Wm. Fountleroy Carpenter probably supported the Jay Treaty

---

* The following appears in the Gentleman's Magazine for August, 1796.

"Died, aged 21, William Fountleroy Carpenter, son of a Devonshire Gentleman, who removed to Essex County, Virginia, where Mr. Carpenter was born, and resided with his mother and three brothers. He had recently come from America to get an estate that had been left him by Coryndon Carpenter, of Launcester, Cornwall. He was unfortunately killed in a duel which he fought with John Pride, a native of Virginia also, aged about 25.

"No previous animosity subsisted between them till the meeting took place, in consequence of a conversation at the Virginia Coffee House on Friday last, in which his ardor upon political topics induced him to reprobate the principles of some of the Congress, who oppose the treaty lately concluded between this country and America. His antagonist was equally warm against those who stood up for the treaty. Very early on Sunday morning they met in Hyde Park, attended by three seconds, who used every means in their power to bring the affair to an amicable adjustment, but in vain.

"The distance of only five paces being measured, they fired at exactly the same instant, when Mr. Carpenter received his antagonist's ball in the side, which penetrated nearly through his body, and, notwithstanding it was immediately extracted, he died the next day at Richardson's Hotel, Covent Garden, and the coroner's inquest rendered a verdict of willful murder. Mr. Carpenter behaved with the greatest composure; remained sensible to the last, and died without a struggle. His last wish was that neither his antagonist nor the seconds should be prosecuted. He was an uncommonly fine young man."

of 1795, because it contained a provision that American claims to property in England should not be held invalid by reason of the war of Independence. In other words, the rights of Americans were to exist as if there had been no separation from England.—Ed.]

At the beginning of the present century the heirs of Lord George Carpenter, of Ireland, along with many others, made claim to a fund in chancery, in a cause known as "The Shadwell Water Works," but they were unable to establish a claim.

Lord George Carpenter died about the year 1731; his grandson was afterwards created Earl of Tyrconnell, and the family is now represented by the Marquis of Waterford and the Earl of Shrewsbury.

It is an interesting case, and illustrates the manner in which chancery cases are dealt with. I believe the original suit is dated more than one hundred and fifty years back. The fund has been recovered in 1872 for Lord Rivers and Sir Frederick Henry Bathurst, who succeeded in making out their claims.

What I wanted was to examine the books personally, and as I found that nothing could be done toward getting to examine the books myself, I set about ascertaining their contents by another method. It was but reasonable to suppose that some time or other, upon some occasion or other, some officer or other had caused an examination and report to be made of the money in the chancery funds.

Pursuing that idea, I discovered that such examinations and reports had been frequently made.

I procured them all, and have them here.

The first was made upon the 5th of April, 1819, at which date the House of Commons had ordered "A return of the total amount of the effects of Suitors in the High Court of Chancery, in the years from 1756 to 1818, inclusive.

This return did not give names and amounts specifically, but if the William Carpenter fund had been swelling for more than a century by accretions of interest, as was claimed, it must have amounted to several millions at an early date.

And yet the return shows, that in 1756, the total amount on hand in CHANCERY was only £2,865,975, 16s., 1d, and down to 1806 the total amount had only reached £21,922,754, 12s, 8d. When it is considered that these amounts included all the money belonging to Chancery funds of every description, all funds of wards, suitors, fees, trusts, etc., etc., it will be at once apparent that no William Carpenter money was amongst it, unless to a very inconsiderable amount. Again : The House of Commons, February 10th, 1829, ordered another similar return to be made up to 1828.

The total amount then on hand was £39,216,526 0s, 1d.

Then came another more searching order, made by the House of Commons February 9th, 1830, ordering a return which should show explicitly the whole amount then on hand.

By this time (1830) the Carpenter fund must have amounted to at least £30,000,000 sterling, which is an amount less than any estimate yet placed upon it.

But that return shows that at that date the grand total from all the above sources then on hand, was :

In cash, £1,496,337, 4s, 2d.

Securities, £37,719,988, 15s, 11d.

Making a total of less than forty millions sterling.

This destroyed even the probabilities of such a fund having been in existence ; but still I have not got at the books to see in whose names the funds stood.

If I could have inspected these it would at once have settled the question as to whether any part of the money was derived from or belonging to the heirs of Wm. Carpenter. So I again applied to the *Accountant's Office* for permission to make a personal inspection of the books, but was refused. I succeeded, however, in getting official copies of all the returns made to the House of Commons, and submit them to you.

They are very instructive.

In 1841 the House of Commons appointed a SELECT COMMITTEE to report upon the proposal to erect new buildings for the Courts of Law and Equity.

The committee had before it the Lord High Chancellor, several of the JUDGES of the HIGHER COURTS of ENGLAND, and many of the most eminent lawyers in London, who delivered their opinions under oath.

In 1842 the report of the COMMITTEE was printed.

It appears from that report, that in 1841 the total amount of stock and cash then standing in the name of the Accountant-General was :

Stock, £39,192,210, 7s, 1d.

Cash, £1,759,629, 2s, 9d.

Among other witnesses before the Committee was the distinguished LORD LANGDALE, who, in discussing whether the funds then on hand could be used towards erecting the proposed buildings, gave a full history of all the funds.

I make the following extract from his testimony.

"It seems an amazing thing to say that there are not less than £41,000,000 or £42,000,000 of stock standing in the name of the Accountant-General, but of that sum there are about £30,000,000 standing in the name of the SUITORS of the COURTS, liable at any time to be asked for by the persons who have a right to their respective portions, as soon as the questions affecting them are decided, and the time for payment arrived.

I suppose nobody will be found to propose touching any portion of that large part of the whole fund standing in the name of the Accountant-General.

Another part of that fund consists of some accumulations made from what is called the SUITORS' FEE FUND, amounting, according to the last return which I have seen, to between £66,000 and £67,000, but which has no doubt been considerably increased by subsequent investments.

That part of the fund arises from the investments of fees received from the suitors; and it is declared by Parliament to be a guarantee for any deficiencies which may happen in the receipt of fees for the payment of certain compensation, salaries and expenses; nobody, therefore, will propose to touch that part of the general fund.

There remains, however, about two millions and a half 3 per cent.

stock, which is commonly called the SUITORS' FUND, and which consists of two distinct parts: one part of it is that which has arisen from the investments of suitors' cash in the Bank of England, and amounts to about £1,574,000. The other part of it consists of investments and accumulations made from the surplus interest arising from those investments of cash, and amounts to about £937,000.

The two together amount to two millions and a half, and a little more. Now, the cash standing on the books as due to the SUITORS of the Court of Chancery, according to the last return which I had seen, amounted to £1,445,529.

The amount or balance of cash standing to the SUITORS' CREDIT in the Bank of England at this same time, viz., in Oct., 1840, was £145,642. . . . .

The stock purchased with the £1,300,000 cash, amounts to about £1,574,000 3 per cents.; and this, for the sake of distinction, may be called he CASH FUND, as arising from the investment of suitors' cash.

It is a security for the payment of so much of the cash as shall be required.

Whether the security is sufficient for payment of the whole which may possibly be required, is a question to be considered with reference to the prices at which 3 per cents. must be sold, to produce £1,3000,000 cash.

It stands thus: the dividends of CASH FUND amount to about £17,000 a year; the dividends of the SURPLUS INTEREST FUND amount to about £28,000 a year; together, they may be said to produce £75,000 a year, and the charges, as appears by the RETURN now shown to me, amounted, in the year 1840, to about £52,000.

Of this sum, the salaries of the Lord Chancellor and Masters amount to £35,000.

There are various other expenses, some certain and some uncertain, which raise the charge upon the income of £75,000 a year, to the sum I have mentioned, about £52,000. The surplus, therefore, is the annual sum of £23,000.

This explicit statement further established the fact that there was no such fund, in amount at least, as claimed.

After getting this, I found another source of information equally definite and explicit.

I found it an Act of Parliament, which was passed in 1853, called "The SUITORS' FURTHER RELIEF ACT." Under this Act the Lord Chancellor was to cause an investigation to be made as to all the accounts whatsoever then standing in the name of the Accountant-General, to the CREDIT of any cause or matter in CHANCERY.

If the accrued dividends had not been dealt with for fifteen years or upwards, he might order them to be carried into a new account to be opened, and called "THE SUITORS' UNCLAIMED DIVIDEND ACCOUNT," and the surplus cash accumulating from this account might be carried to the credit of "The Suitors' Fee Fund Account."

In 1854 the first investigation was made under this Act. It then appeared that the total amount of SUITORS' STOCK then in COURT was £46,000,000.

In 1855 a list of these accounts was published, but no names or amounts were given.

In 1860 and 1866 similar lists were published, but there were no names or amounts given in them, so that I was unable to say definitely whether or not they included the William Carpenter fund. I could only say that, judging from the sum total, it was not large enough to indicate any such fund.

Subsequently I found in the British Museum a work which did give the names and dates. It is "The Unclaimed Dividend Book of England," containing the names and descriptions of upward of 20,000 persons entitled to various sums of money, being the whole of the unclaimed dividends and stock in the "Public Funds," which have been, by the terms of the "Bank Charter," transferred to the Commissioners for the Reduction of the National Debt, as unclaimed for ten years and upward.

Of course I examined this with the utmost care, but, as usual, I found nothing in it to the credit of any Carpenter.

Next in point of time was a list, which was the most satisfactory of all. This was an official list published in 1847.

It gives the names of the suitors and title of the causes wherein they are engaged.

It covered every point of inquiry, as will be seen by reference to the heading of the list, which is as follows :

"A list of the titles of causes, matters and accounts in the books of the *Chancery Pay Office*, to the credit of which funds were standing on the 1st September, 1875, which had not been dealt with during the fifteen years immediately preceding that date, prepared pursuant to 'Rule 91 of the Chancery Funds consolidated Rules, 1874.'"

It was in reference to this list that the rule was established, restricting information, which I have mentioned heretofore.

Although ostensibly confined to causes not dealt with for the preceding fifteen years, in reality it includes all causes and matters up to that date.

It covers sixty-two columns and embraces over three thousand names; and if the Carpenter fund was a reality and not a myth, it would certainly appear there. It could not be omitted.

But you will see that there is no suit whatever relating to any administrator of any Carpenter estate, or Carpenter trust, although, by inspection, it will further be seen that the list deals with just this class of cases.

As witness the numerous suits instituted in the Freehold estate, and inheritance of "certain pieces of ground," and the others as numerous concerning the "Capital Account of the person or persons entitled to shares of stock, standing in the name of, etc., etc., and in other matters of trusts, etc.

"THE RECORD OFFICE."

I am now going to make excavations in those venerable and time-out-of-mind documents which are kept in the Public Record Office, and where some of the valuable materials are so ancient that they almost crumble into dust as one looks at them through a strong magnifying-glass.

The oldest existing English records are tallies in exchequer, which, down to 1834, continued to be used both for receipts and for simple records of matters of accounts.

They consist of wooden rods, marked on one side with notches, to indicate the sum for which the tally was an acknowledgment, while on the other two sides were written the amount, the name of the payer, and the date of the transaction, and the tally being divided longitudinally, the one-half was preserved in exchequer, and the other given to the person who had paid the money. This rude contrivance, which came down from *Anglo-Saxon* times, was an effectual safeguard against forgery. Parchment is the material on which the greater portion of the records are written, the skins being attached at the top book ways.

There are two reading-rooms filled up for the use of the literary and antiquarian public. The interior building is of rather an heraldic order, as coats of arms adorn every visible part which is not filled with ancient and vellum-clad indexes.

There are seldom more than a dozen readers here, and those are strictly of the antediluvian species. Their garments redolent with the dust of ages. Now and then you see one of the gentler sex there, but not often, and those, as a rule, are of the spectacled and strong-minded stamp, well up in the dead languages. In visiting the " Record Office " I had this object in view: I was not entirely satisfied with the denial at the Chancery Office, and I wanted to get at other sources of information. It occurred to me that when the Accountant-General received money, or paid it out, a receipt must have been given for it, and these receipts should be on file, and would show every transaction. It proved that I was correct in this supposition, and the receipts were filed in the Record Office. I found the original certificates of the Accountant-General's from 1695 to 1799, packed away in two hundred and ten bundles, in the Record Office, unnumbered, and without index. I spent nearly one day in having one of these bundles exhumed. It was for the year 1720; I selected that year, because, if any fund had accrued, it had accrued prior to that date, but, as usual, the search was fruitless. There was no receipt in the bundle for any Wm. Carpenter money of any sort. I intended to search the remaining bundles, but was told by the chief clerk that it might take months to get them out and examine them, and that it was unnecessary to proceed further in that office, because there was a complete index in the office of the Accountant-General, which could be consulted.

Accordingly, I again went to that office, and asked to have the index examined; the clerk examined the same, and stated that there was positively no receipt for any William Carpenter money. Not being there, the next query was, where else could it possibly be ? This seemed an insoluble conundrum. But, finally, it occurred to me that possibly the estate might have included lands as well as money, and that, for want of heirs, both had been taken by the crown upon escheat. Determined to follow up every chance, I made long search through the records at the record office, and patiently ploughed through the labyrinth of these old and rugged paths in exploring these musty old parchments. Most of the ancient documents are in Latin, which is not classical, not ungrammatical, but abounding with words which are unknown in the school, and is abbreviated, and often to such an extent that a single letter represents a word, and requires much time and patience before one can make head or tail of them. I was, however, fortunate enough to find a complete record of escheats, which soon dispelled all hopes in that direction. I found the record, the title of which is:

"Inquisitionem, Post Mortem Sive Escheatorum."

It contains a list of all lands, estates and funds which, for want of heirs, escheated to the crown. It begins I. Edward I., A. D. 1274, and is brought down to November 29, 1826. It shows that there never was an escheat of any William Carpenter estate, so that prop was gone also.

SOMERSET HOUSE.

Having left the Record Office, the first step it behooves me to take will be to shake off the dust of ages which covered my clothes during my researches among the venerable and crumbling records mentioned in the last chapter, and as it was stated that Wm. Carpenter left a will, I betook myself to the Somerset House.

Before paying the fees and commencing the search, I will take a glance at the place itself and all its proceedings, as a brief description will, I think, be interesting to the reader.

It is situated in the Strand, Doctors' Commons, and stands on the site of a palace built by Protector Somerset, about 1549, and which escheated to the Crown on Somerset's execution.

The original Somerset House was torn down and rebuilt in 1776, and was built in the Palladiain style for public offices; various offices connected with public departments are in the building, in the east wing of which is the King's College.

On going in everything seems hurry and confusion.

Rapidly from top to bottom of pages run the fingers of the barristers, who are dressed in long black gowns and white wigs, which gives the place and its surroundings a somewhat ancient aspect. They turn folio after folio of the bulky volumes which they are examining, long practice having taught them to discover at a glance the object of their search. Rapidly glide the pens of the numerous copyists who are transcribing or making extracts from wills.

But as we begin to take in a little more clearly the busy throng, occupying the central space we see persons whose appearance and manners exhibit a striking difference to those around them. There is no mistaking them as solicitors, or solicitors' clerks, acting for other persons; but as acting for themselves, looking for their own interest.

There are few places where the careful observer of characters has so many opportunities of watching the various passions and feelings of the human race as in the search room at Doctors' Commons. So absorbed becomes the searcher in the business that for a time he forgets all else.

You see a face exulting with hope and expectations gradually grow darker and darker as the contents of a will is revealed; or your eye rests on a beggar in gait and attire, hardly able to read, spelling over every word, and mumbling to himself all the while. In fact, one might almost write a novel from what the imagination will conjure up from the varied expressions and transformations the faces undergo as they take in the substance of the will.

Over some will pass gleams of vicious delight to see that some one had been cut off with a shilling, though they themselves have benefited nothing; over another face will steal an agony of expression on discovering that they have been forgotten or neglected, when, perhaps, they had centred all their hopes on some expected legacy to free them from debts, which, like millstones round their necks, were dragging them down to beggary.

Then we see the inveterate fortune-hunter, who has searched the wills of his or her ancestors scores of times, in the vain hope that there is some clause hitherto not observed, which will enable them to make a claim. Again, there is the imaginary claimant, *generally of the female sex.* She has her pedigree by heart, has certificates by the score, but there is always some date of marriage, birth or death that stands between her and fortune. On and on she works, getting thinner from disappointed hopes, more aged than her years, and with gray hairs prematurely round a face which might have worn once a happy, contented look, but which, by the constant wear and tear of anxiety, has now a discontented, querulous expression.

Many of these characters may be seen in Doctors' Commons; another not uncommon case is the disappointed relative. A relative has lately died well-off, and has always promised to do something for John, Thomas or Kate. They hear nothing of the expected legacy, and have a vague idea that by going to Doctors' Commons, and paying their shilling, they will see the will, and all they will have to do will be to apply to the executors and get their money. With what eager eyes the big book is searched; remarks, perhaps, are heard about "the poor dear old man, how quietly he went off at last." "What a blessing he did not suffer more," etc.

Gradually, as the end of the will is reached, the changed expression of the face is marvelous. The regrets are changed to "Stingy old beggar;" "the shabby old fellow," etc.

A many fast looking and highly got up young man may always be found here, looking very much out of place, and seeming in a very nervous state of mind. They are penniless Government clerks looking out for rich wives to support and keep them in luxury; they go there to see what the grandfather or the father (if he be dead) has left to some fair Julia or dazzling Rosalind, who has bewitched him, but to whom he cannot afford to pay his devotions unless they are set in gilded frames.

Having explained the nature of the proceedings in this court, we will now begin the search for the "will" of Wm. Carpenter; we must purchase a shilling stamp, which stamp, when procured, we take to the little box on our right as we enter the room, and hand it over to the very urbane official there, whose sedentary life and little occupation seems to agree with him vastly well, from the sleek appearance he rejoices in. This well-fed worthy asks us, "What name?" We give the noble patronymic, which he writes on a paper, and with a stereotyped official grin bids us to take it to No. 6. We accordingly hand the said paper to a deeply-tinted golden-haired clerk, who says, "What year?" Upon which we give him a number in round figures, liking plenty of sea-room. This ruby-crowned gentleman points out where the indexes live, and we proceed to search in those containing the desired years, and carry them one at a time to the centre desk; we stood among the anxious throng there, soon got absorbed and excited in our chase; we searched for over a period of fifty years for any will or letters of administration, but could find no record of either.

THE STATUTE OF LIMITATION.

At this time my attention was called to the Statute of Limitation, which was recently passed, and which put an entirely new phase on the whole matter. Previous to this time only the ordinary Statute of Limitation applied to such claims.

The following are some extracts from An Act for Limitation of actions and suits relating to Real Property, and for simplifying the Remedies for trying the Rights thereto.

SEC. II. "No person shall make an entry or distress, or bring an action to recover any land or rent, but within twenty years next after the time at which the right to make such entry or distress or to bring an action, shall have first accrued to some person through whom he claims, or if such right shall not have accrued to any person through whom he claims, then, within twenty years next after the time at which the right to make such entry or distress, or to bring such action, shall have first accrued to the person making or bringing the same."

SEC. XII. "Provides that any person under the following named disabilities, viz., infancy, coverture, idiocy, lunacy, unsoundness of mind, or absence beyond sea; then such persons claiming through him may, notwithstanding the period of twenty years hereinbefore limited shall have expired, make an entry or distress, or bring an action to recover such land or rent at any time within ten years next after the time at which the person to whom such right shall first have accrued as aforesaid, shall have ceased to be under any such disability, or shall have died (which shall have first happened)."

SEC. XVIII. "Provides that no entry, distress or action shall be made or brought by any person but within forty years next after the time at which such right shall have first accrued. Although the time at which such right shall have first accrued, although the person under disability at such a time may have remained under one or more of such disabilities during the whole of such forty years, or although the term of ten years from the time at which he shall have ceased to be under any such disability, or have died, shall have not expired."

These sections are the law in England to-day, and even if it proved to be true that William Carpenter died and left property, and if more than forty years has elapsed since any of his descendants first had a right to the same, whether they were ignorant of the fact or not, their remedy is absolutely gone.

Here follows a late enactment, which showed me that even if I had discovered the fund, and if it consisted of personality, as was stated, it could not have been recovered, except under certain circumstances, which I shall hereafter explain.

Sec. 40, chap. 27: " To cases of claims to estates of persons dying intestate previous to this time, only the ordinary limitation applied to claims to *personal property.*"

I have been informed that this section was introduced for the declared purpose of preventing the frauds which were constantly perpetrated under pretense of reclaiming vast estates, and was particularly designed to terminate all further efforts which were being made at that date to reclaim an estate which was and is better known than the famous Jennings estate.

After careful examination, I expressed doubts as to its constitutionality being retroactive.

I was reminded, however, that there was no written constitution in England, that Parliament was omnipotent, and that no matter what might be its history or effect, the statute had been acquiesced in and repeatedly acted upon, and no court in England would now venture to disturb it.

No suit or other proceedings shall be brought to recover the personal estate, or any share of the personal estate, of any person dying intestate, possessed by the legal personal represent,tives of such intestate, but within twenty years next after a present right to receive the same shall have accrued to some person capable of giving a discharge for or release of the same, unless in the meantime some part of such estate, or share, or some interest in respect thereof, shall have been accounted for or paid, or some acknowledgment of the rights thereto shall have been given in writing, signed by the person accountable for the same, or his agent, to the person entitled thereto, or his agent, and in such case no action or suit shall be brought but within twenty years after such accounting, payment or acknowledgment, or the last thereof, if more than one was made or given.

The certain circumstances to which I referred as being the only ones which could avoid the above statute, were:

1. That circumstances could be discovered of such character that a trust could be established.

2. That circumstances could be discovered upon which it could be averred that a fraud had been perpetrated upon the claimants, and the knowledge of such fraud concealed from them to the present time.

I thought it would be well ere leaving London to visit Mr. Phillipps, the person who claimed to possess so much valuable information.

I made my way to No. 93 Highgate Road to see this person, but was informed that he had removed to 518 Caledonia Street. From slight intimations which I had accidentally overheard, I concluded to make some inquiries about him.

Meeting a lady friend I inquired into the matter and she seemed to know him very well.

"He is a very eccentric and odd person," she said, shaking her head, and tapped her forehead a great many times with her forefinger, to express to me that I would find him such; "for he is a little —M——you know. Why," said she, " his name is not Phillipps, that is an alias; his proper name is Harrison."

On calling at 518 Caledonia Street, a hard featured female, with a red nose and rather unsteady eye, but smiling all over, opened the door after I had tapped.

She planted herself in the doorway, as if she was about to crack herself like a nut. This was Mrs. P., who has a remarkable talent for silence.

She was very tart and spare in her replies, these sufficing, however, to inform me that Mr. P. could be seen at his office. She would not say whether Mr. P. was in the house or not; the only satisfaction to be obtained was, I could see him at his office in the morning.

Bright and early next morning I called at No. 10 New Court, Lincoln's Inn. Mr. P.'s office is squeezed up in a corner. Three feet of knotty floored dark passage brings you to the entrance of Mr. P.'s office, in an angle profoundly dark on the brightest midsummer morning.

Mr. P.'s office, to give a correct illustration, is on so small a scale, that his clerk can open the door without getting off his stool, and has equal facilities for poking the fire.

The place from appearance was last painted beyond the memory of man. The dull crooked windows in their heavy frames have but one piece of character in them, which is a determination to be always dirty.

" Walk in the back office, sir," said the clerk, upon entering. Mr. P. sits facing round on a stool at the desk, turning towards the door. He is tall and withered, with his head sunk sideways between his shoulders, and the breath issuing is visible smoke from his mouth, as if he were on fire within. His face is stern and was much flushed. If he were really not in the habit of taking rather more than was exactly good for him, he might have brought an action against his countenance for libel, and have recovered heavy damages.

"Is this Mr. P.?" I asked on entering this back room. "Yes, sir." He coughs behind his hand modestly, anticipating profit. A long conversation then ensued; he answering me slowly and evidently with reluctance. He is accustomed to cough with a variety of expressions, and so to save words. He kept his eyes upon a note book-I held in my hand.

He confessed during my interview, that he knew nothing about the *estate*. I reminded him of the letter he had sent me, in which he stated that Wm. Carpenter died in 1700, and the government alleged that he was an illegitimate.

He said he knew nothing of any such letter, and seemed surprised and embarrassed when I produced the same. He then said his intentions were to get up the genealogy only.

I told him that he laid himself liable to the law, for obtaining money under false pretense. I asked him what evidence he had to show that Wm. Carpenter died in 1700, and that there were some proceedings which had taken place in Court regarding his property, when he knew that there were no such records.

He said that some person whom he met at the Record office, but whose name he had forgotten told him so.

"Why," said he, drawing himself to his full length, "you people from America said so."

He openly brags that he would do nothing for an American without having the money in hand.

He confidently expects to receive an amount from Mr. Barker to open "*fire*" as he calls it. He became quite confidential, and said that in his opinion the American people were all fools on the subject of genealogy.

As I got up to start, he laid his hand upon my shoulder and said with a sickly smile, "Always here, sir. Personally or by letter you will always find me here, sir, with my shoulder to the wheel."

Thus we part, and as I emerge from the heavy shade of Lincoln's Inn, into the sunshine of Chancery Lane, for there happened to be sunshine there that day, it seemed to me that when a man's conscience begins to get hard, it becomes so faster than anything in nature. It is like the boiling of an egg; it is very clear at first, but as soon as it gets cloudy, one minute more and you may cut it with a knife.

PRICKING GOLDEN BUBBLES.

I will now endeavor to show the reader how successful this business is carried on in England.

While there I met three American gentleman who had recently arrived to lay claim to an estate valued at three hundred millions of dollars.

Upon my arrival here, almost the first newspaper I saw my eyes crossed an announcement concerning a *William Bradford Estate*—another *phantom estate*—about equal in amount and similar in circumstances to yours and other estates.

It is evident at once that the sums claimed for recovery in these cases would alone bankrupt any nation on earth.

The number of those who have great expectations is immense. Always waiting for his or her ship to come in. It is expected from all quarters. It is a rich uncle or aunt or some relative who is sure to leave his money to us when he dies.

The most of these hopes are but pleasant dreams that do no harm and may lighten the burdens of the present by making us hopeful of the future, but the great expectations that are founded upon supposed heirships to great unclaimed estates in England are not only delusive but dangerous.

There are thousands of people who honestly believe themselves to be directly or indirectly entitled to vast sums in foreign countries that are said to be awaiting only proof of ownership. These estates are almost invariably mythical.

They are kept alive mainly by dishonest claim lawyers who prey upon the credulity of would-be-heirs, and so gather from them sums which they allege are to be expended in the prosecution of claims which are really unfounded.

These vast estates abroad are ships that never leave their mooring. Rather than wait for them to come in we had better go to work and build a ship for ourselves.

The fruit of our own industry is a ship whose swelling sails always bring it nearer.

The following is a list of *phantom estates* which have been a source of revenue for claim lawyers for many years, with a table showing their *supposed* value and the number of heirs among whom they will be divided *when recovered* :

	Heirs.	Estate.
Anneke Jans,	1,000	$317,000,000
Baker,	87	250,000,000
Sir Hugh Mosher,	—-	200,000,000
Chadwick,	5	37,000,000
Edwards,	160	9),000,000
Joseph Wilson Ingraham,	—	500,000 000
Hyde, N. G.,	200	12,000,000
Hyde, Ann,	150	360,000,000
Hyde, Brooklyn,	1	5,000,000
Jennings,	1,835	400,000,000
Hedges,	—	250,000,000
Kern,	100	200,000,000
Leak,	—	100,000,000
Mackey,	1	10,000,000
Merritt,	80	15,000,000
Shepherd,	15	175,000,000
Trotter,	200	2 0,000,000
Townley-Chase,	—-	1,800,0,0,000
Lawrence-Townley,	1,000	500,000,000
Van Horn,	20	4,000,000
Webber,	60	50,000,000
Weiss,	4	20,0,10,000

Grand total : 22 estates, 4,918 heirs. Value of estates, $5,490,-500,000.

In one of these cases (the Hyde case), I am informed, several thousand dollars were spent in a vain endeavor to try and find the location of the estate.

When I became convinced that there was no money to be reclaimed, as I did at an early date, I felt that it would be at least satisfactory if I enabled them to trace with accuracy the pedigree of their ancestors.

For this purpose I collected extensively, but without arriving at any definite result.

What I have collected, however, I present to the Association.

The records of Carpenters which are preserved relate almost wholly to the titled families and those known as "the gentry." It was a safe assumption from the first that your William Carpenter was not included in them.

The term "gentleman" formerly and now, in England, does not have only the signification given to it in America ; it has a meaning which enters into and forms part of the social, landed and political organization of the country. But as I was searching for a possible identity or connection between the Carpenters who were entitled to dividends on annuities and those who were advertised for in the list heretofore given, and the Carpenters through whom you hoped to claim, it became necessary to enlarge my search as far as possible. It occurred to me that there might be in existence a book containing the names of all who left their country, with the date and such other particulars as would naturally appear in connection with it, and which could give me much desired information.

Therefore I went, first, to the Record Office. I collected all there was in that office of the name of Carpenter, which, however, was of little service.

The next place which suggested itself to me at which to look was "Lloyds' Shipping Register," at the Royal Exchange. Accordingly I went, but found there was nothing definite to be obtained there. In 1838 all their records had been burned in the fire which destroyed the Exchange.

I next went to the "Emigration Office," a Government institution in Dock House on Little Tower Hill, but obtained nothing there, as their records were too modern to be serviceable.

A similar result followed at the Board of Trade.

"THE BRITISH MUSEUM."

An important national institution. It originated in a bequest of Sir Hans Sloane, who, during a long lifetime, gathered an extensive, and, at that time, unequaled collection of objects of natural history, besides a considerable library of books and manuscripts. These, in terms of his will, were offered in 1753 to the Government, on condition that £20,000 should be paid to his family, the first cost of the whole having amounted to more than £50,000. The offer was accepted, and the collection, along with the *Harleian* and *Cottonian* libraries, were arranged in Montague House, which had been purchased for £10,000. The new institution, thenceforth called British Museum, was opened in 1759. On each side of the Museum there is a semi-detached house, containing the residence of the chief officers of the establishment. The grand entrance-hall is a noble and lofty apartment, built in the massive Doric style. It contains a statue of *Sir Joseph Banks* and an ideal representation of *Shakespeare*.

The interior of the building is admirably adapted to the purpose for which it is devoted; some of the galleries, from their size and dimensions, have a very imposing appearance, as the king's libra y. etc.

The rea ling-room is circular. It is constructed principally of iron. with brick arches between the main ribs; the dome is 106 feet in height, and its diameter 140 feet, being second only to the *Pantheon of Rome*, and that but by two feet.

There are over three miles of lineal of book-cases eight feet high, and range about twenty-five miles of shelves. Assuming these shelves to be filled with books of average size, the leaves placed edge to edge would extend about 25,000 miles, or more than three times the diameter of the globe.

The reading-room contains over 700,000 volumes, which are not arranged with reference to subject matter, but are classified by authors and titles. When the title of a book and its author is known, the book can be obtained at once, but when neither are known, but it is desired to search promiscuously in the vague hope of procuring information, it becomes almost a matter of impossibility to proceed, inasmuch as the rule requires that the name of the author, title, place, date and size of the book wanted shall be stated, and if a manuscript, its presswork also. Only one volume at a time shall be asked for on a ticket.

Determined to find something which would throw light upon the question as to where and when your ancestors left England, if there was anything which would do it, I made long search, and at last was fortunate enough to find all there was in existence, gathered together in a volume which, it is true, only come down to A. D. 1700. But as the statements of the editor were as applicable to later as to prior date, I was compelled by them to forego any further expectation of finding what I wished.

The book was entitled:

"The original lists of persons of quality, emigrants, religious exiles, political rebels, serving men sold for a term of years, apprentices, children stolen, maidens pressed, and others, who went from Great Britain to the American Plantation A. D. 1600–1700, with their ages, localities, where they formerly lived in the mother country, the names of the ships in which they embarked, and other interesting particulars, from MSS. preserved in the 'State Paper Department of Her Majesty's Public Record Office.'"

The title of the book included every point of my inquiry, but the book itself did not fulfill the promise of the title. In the preface the editor says:

"It cannot be doubted but that other lists were made, but they are either lost or are among the mass of papers still uncatalogued at the Record Office. We learn incidentally that ships left England almost daily for America, but no record of them or their passengers remain. I know that many ships sailed from Bristol, among others the '*Angel Gabriel*' and the '*James*,' conveying the Rev. Richard Mather and the Rev. Daniel Maud."

I therefore concluded to visit Bristol, expecting to find records concerning the sailing of vessels from that port, with a list of their passengers. Upon my arrival at Bristol I found that such lists had been carefully made and kept on file in the Custom House; but, unfortunately for me, these records were or had been destroyed by a mob of rioters in 1831—they having destroyed the Custom House, with other buildings, by fire.

The next place which suggested itself, was the Merchants' Commercial Association, where there was a record for two hundred years of the sailing of vessels, but none of any "Carpenter." Then the librarian of the Public Library, who is an antiquarian, but he could give no information. Among the thousand who emigrated, it cannot be doubted but that a very large number left to avoid the payment of subsidies, and that they would not take the oath of allegiance and supremacy. These, therefore, must have left secretly, and of such no record would remain.

As I have said, these were conclusive arguments equally as potent after 1700 as before it, and as no other volume has been published since on the same subject, it satisfied me that there was not any accessible records to tell the date of your Ephraim, Josias and Timothy's departure, unless it could be found in the register of the parish where they were born. But where were they born? Not knowing that, I might have searched all the parish records in England fruitlessly. The preface, however, of Mr. Hotton's book did corroborate the fact as to William Carpenter, age 62; William Carpenter, Jr., age 33: and Abigail Carpenter, age 32, of Harwell, who left Hampton, May, 1638, on board the *"Bevis."*

Among the many persons whom I called upon, was Mr. A. B. Carpenter, of Temple Court, London. He had previously heard of the published reports, but thought that the £40,000,000 was but an Arabian Knight's Entertainment, and the whole story was too absurd to think about. His grandfather was the Mayor of Sutherland, and his father, Charles Carpenter, the youngest son of a large family, was a clergyman, and resided at Alresford Hants.

Messrs. William Carpenter & Son, Attorneys of Pountney Lane, London, stated that they knew nothing of any such fund, and said that they did not believe that there was anything of the sort to be got, and that they would much prefer to work for whatever money they got, rather than await for any from such an unlikely source.

I learned that a gentleman residing in Manchester could give valuable information, and consequently a journey to Manchester was made to see him. Mr. Richard Carpenter, of 27 Brazennose Street. The family formerly resided in Dublin and Galway; his father, John, who died in Manchester, 1878, had two brothers; Thomas, the eldest, died in Galway, in 1882, and William died in Panzants, Cornwall, in 1866. The family were interested in, and had devoted much time in tracing their pedigree, in order to recover a share of the property which was left to Dr. Nathaniel Carpenter, who died in 1778, in Virginia, amounting to about £3,000.

They have never taken the matter into court, and when I explained about it being reported that the Manchester Carpenters endeavor to recover the estate of William Carpenter, he stated that his family was the only one of that name in Manchester at that date (1846), and that they certainly never made an effort to obtain it, and he had never heard of the claim before; furthermore, there was no Henry Carpenter residing in Manchester at that date. "Prince's Worthies" mentions that John Carpenter was a Cornish gentleman, and died in 1620, leaving a son, Nathaniel, who graduated at Oxford College. The said Nathaniel was born in 1588 and died in 1643. He became a dean in Ireland, and was celebrated as a "philosopher, mathematician and poet;" his biographer states that he proved himself a Nathaniel—born of God, and a Carpenter, a wise builder of God's Temple. It is probable that the Carpenter family of Ireland are descended from him, and they are related to the Carpenters of Cornwall. I have greatly abbreviated the details of the foregoing history.

In working the Carpenter pedigree out, several other branches of the Carpenters were incidentally traced to their roots, and I give a synopsis, not as bearing in your line at all, but as a matter of interest, which those who have the time and curiosity may elaborate if they will.

CARPENTERS OF MOUNT TAVY:

John Carpenter, Esq., of Mount Tavy, Co. Doven, born Feb'y 28, 1839; S., his father, May 16, 1842. Lineage:

John Carpenter, Esq., who m. Mary, sister of William Spry, of Tavistock, had (with four daughters, viz: Hannah, m. to John Bolt, Esq.; Mary, m. to Rev. C. Porter; Deborah and Catherine) six sons:

John, father of Mrs. Hele, of Kingston House.

Nathaniel, who had a son, Coryton, and two daughters, one m. to Admiral Badger, the other to Mr. Rowe, of Lanceston.

Samuel, of Lanceston, who m. Eliz. Hodge, and had with two daughters four sons. 1. John, whose son John, an officer in the army, m. in 1797, Teresa, daughter of George Fieski Heneoge, Esq., of Hamilton Co., Lincoln, and had issue. 2. Charles, of Maditonham. 3. James, Admiral R. N., m. and had issue. 4. Samuel, Barrister at Law, d. in 1815, leaving issue: Joseph, father of the Rev. Wm. Carpenter; Benj., of whom presently; Phillip. The 5th son, Benj. Carpenter, m. Patience Edgecomb, and was father of John Carpenter, Esq., who m. Christian Phillipps, and had issue: John Phillipps, his heir, Benj. Edgecomb, Charles Coryton, George, Eliz. Pomeroy, Patience, and Christian. The eldest son, John Phillipps Carpenter, Esq., of Mount Tavy, Co. Doven, m. Eliz. Stubling, and d. in 1812, leaving (with a daughter, Patience—Christian, m. Aug. 24, 1807, to Sir Wm. L. Salisbury Trelavey, Bart. of Trelavey) two sons: John, his heir, and Jonathan—Phillipps in hold order, who m. April 16, 1827, Harriett, oldest daughter of the Rev. William Garner, and d. Aug. 26, 1841. The eldest son, John—Phillipps Carpenter, Esq., of Mount Tavy, J. P., b. Dec. 30, 1796; m. July 19, 1826, Lucy, fourth daughter of the Rev. Wm. and Lady Harriet Garner, of Rooksbury Park, Hampshire, and d. May 16, 1842, leaving (with three daughters, Eliz.—Harriet, m. 1852, to the Rev. H. M. Sims, rector of Hinderwell; Lucy, m. 1852, to Henry Clark, Esq., of Efford Manor, Co. Doven, and Anne) an only son, now John Carpenter, of Mount Tavy.

The Bible gives in the book of Genesis the genealogies of the patriarchs from Adam to Noah, and from Noah to the twelve patriarchs. In the 26th chapter of Numbers we find the number of all the males of the children of Israel from twenty years old to be 603,550, for Moses and Aaron assembled all the congregation together on the first of the second month, and they declared their pedigrees of their families by the house of their fathers, and in the 7th chapter of Nehemiah we read that Nehemiah obtained permission from Artaxerxes, after the captivity of the Jews in Babylon, to go up to Jerusalem and rebuild the city of the sepulchres of his fathers; after which he relates, My God put into my head to gather together the nobles, and the rulers, and the people, that they might be reckoned by genealogy, and I found a register of the genealogy of them which came up at the first," and this register was of so great authority that some of the priests at Jerusalem sought their register among those that were reckoned by genealogy, but it was not found, therefore were they as polluted, and put from the priesthood.

Parochial registers seem to have begun about the middle of the sixteenth century, shortly after the dissolution of the monasteries, and the dispersion of the monks, who had been until that time the recognized recorders of those events. The bards, or ancient Druids, were much given to composing genealogies and rehearsing them in public assemblies, in which they were very skillful. There have been various opinions as to the precise period when parish registers were first kept in England, but I find in 1538, the 30th Henry VIII., a mandate was issued by the Vicar General, for the keeping of registers of baptisms, marriages and burials, in every parish, before which date there were no parochial registers; the register for some few parishes have entries for two or three years prior to 1538, but there is reason to believe such entries were not made until the institution of registers in that year.

In 1597 it was ordered that certificates, transcripts of the registers, should be then and thenceforth forwarded annually to the registrar of the diocese. In Cromwell's time the registers fell into general disuse, as the established clergy were frequently ejected from their livings, and marriages were celebrated before a Justice of the Peace, and not in the church.

This mischief was rectified at the Restoration, but the evil of having no parochial records for a period of twenty years is incalculable.

The next important regulation was the Marriage Act of 1754, which directed the ceremony to be performed in a church or chapel, and the entry in the register to be subscribed by parties in a proper form. Prior to this Act, a religious ceremony was not necessary to the validity of a marriage, and parties coming together and marrying per verba de presente, were married beyond their power to separate. Although parish registers have been instituted for nearly 300 years, I should be much disappointed if I expected to receive from them the assistance which one ought to obtain, from the negligent way in which they have been kept. The repetition and gross errors in entering them, precludes the possibility of substantiating a pedigree traced through a period of two centuries; for, though furnishing important links of evidence, often when revealed only serve to show how much still remains to be discovered.

Independently of the casualties, especially in the burning of churches during the Commonwealth, much hindrance is caused even to this day, in making and transmitting the transcripts, as well as no proper preservation of the originals, which have been lost, canceled, stolen, left in public courts of justice for examination, and never reclaimed, burnt and even willfully destroyed.

There is another circumstance, the inattention which has too frequently been given to the subject by the party making the entry, as may be instanced in these few extracts from parochial registers.

Baptisms: An infant christened 1570.

The Queen's footman's child, 1554. "Joane Filia Populi."

Marriages: This day were married by Mr. Holloway, I think, a couple whose names I could never learn, for he allowed them to carry away the license. "Inezel, man and maid, was married on Ladyday, 1706."

Burials: "A mayde from the mill."

"Black John."

"Apprentice of Mr. Kilford."

"Goodwife Lee."

"A tinker of Berrye, in Suffolk."

"1716, the ould girl from the workhouse."

"1660, a child of Adam Earth."

"1606, a sucking man-child."

The "Gentleman's Magazine" for 1811, remarks, that in many country places, the clergyman has entered the names at his leisure, whenever he had nothing better to do, and perhaps never entered them at all. Misnomers have occurred in every page, and the registers have often been lent about the parish to any of the friends of the incumbent, or the church wardens, who, from curiosity or worse motives, have been induced to borrow them.

In an Essex parish, the clerk not having ink and paper to make an extract for an applicant observed, "Oh, you may as well have the leaf as it is;" and coolly taking out a pocket-knife gave the applicant the entire two pages of the register.

Bigland, in his "Observations on Parish Registers," mentions in one parish the clerk was a tailor, and had cut out more than sixteen leaves of the old register, in order to supply himself with measures; and in another parish, the register being in the custody of a parish clerk, his daughters, who were lace makers, were allowed to cut up for a supply of parchment to use in their manufacture.

Dr. Thelwall, of New Castle, wrote in 1819, that "The records there were so shamefully kept, as he had seen in the possession of a friend a great number of extracts from the register of a certain parish in the neighborhood, and on questioning how he became possessed of them, was informed they were given to him by his cheesemonger, and that copies where forwarded by the clergyman of the parish to the proper office, in a bordering diocese, and had been allowed, through the negligence of their keeper, to obtain the distinguished honor of wrapping up cheese and bacon;" and in an account of the present state of the Ecclesiastical Court of Record, by W. Downing Bruce, 1851, we find that the parish register of Kirkby Malgeard, Yorkshire, for 1653, was reported by the curate as lost or stolen, and that it was discovered by him (Mr. Bruce) tattered and torn, behind some old drawers in the curate's back kitchen. In addition to the parish churches, there were before the passing of the Marriage Act many chapels in and near London, which exercised the privilege where marriages were licensed to be celebrated, and gave rise to great abuse; of these, the "Fleet" and "May Fair" were the most notorious. May Fair Chapel was the resort of the higher class of society for clandestine marriages, and in those registers (from 1728 to 1754) now kept at the consistorial court of London,

(except three volumes, which are in the church at St. George's, Hanover Sq.) appear amongst other great marriages, that of the Duke of Kingstone, and the celebrated Miss Chudleigh, and of Henry Fox, first Lord Holland, who, in 1744, ran away with the Duke of Richmond's daughter.

Such of the Fleet Registers as could be discovered by Government, were deposited at the Bishop of London Registry in Doctors' Commons. The evidence respecting the marriages in "Fleet" was given by a witness on the trial of Doed Passingham v. Lloyd, was taken down by Mr. Gurnly, and as it is curious, and in all probability quite unknown, I will give a transcript of it verbatim.

Wm. Stiles Jones: I lived in Fleet Lane, I know the houses called "Marriage Houses" and register books were kept at them. The houses extended beyond the rules of the "Fleet." Dr. Dean and Dr. Wyatt were clergymen who celebrated Fleet marriages.

The marriages were set down in a book kept at each of the marriage houses by the persons who acted as clerks.

Mr. Lilly had a marriage house and Mrs. Owens used to ply for him, but not very decently, for she got any one to be married who would. When Lilly died, Owens kept a marriage house on her own account. On his cross-examination he said, "If the clerk was out the servant of the marriage house entered it into the book. Two of the houses were the sign of the Sawyers, and the sign of the Salutation and Cat, in Newgate market, another was the Bull and Garter. Lillys was more of a private house, and had no sign." Benj. Panton, a witness, said "I bought the whole of the registers of "Fleet" marriages, they were between 500 and 600 in number, and are more than one ton in weight."

The manner in which these marriages were celebrated, the conduct of the persons who assumed the power of registering them, and the numerous false entries in them of marriages which never did take place, have thrown such an odium on them, as to take from them even the authority of a private memorandum, although the marriages celebrated in the "Fleet" were undoubtedly valid.

Protestant dissenters of all denominations have been accustomed to register the birth of their children at Dr. Williams' library, Red Cross St., and in some years there have been as many as 1000 entries in each year, but the number has diminished since Sir Thomas Plomer refused to receive the book as evidence.

Since the Act of Parliament in 1832, all registers are transmitted to Somerset House. In London there are three distinct offices where searches can be made. First, the Bishop of London's office, as regards his consistory Court. Secondly, the Vicar General's, which has authority over the whole See of Canterbury, and thirdly, the Faculty office, which has jurisdiction over York as well as Canterbury. The License Records commence in 1630, at the Faculty office, but at the other offices they do not commence until after the fire of London in 1666, all earlier records having been burnt.

In closing my report, I desire to make the following observations.

While my mission has not been productive of pecuniary profit, yet I think the result has been beneficial. It may now be definitely taken as a fixed fact, that the so-called "Carpenter Estate" does not exist, except in the fiction of tradition, and the hopes of the expectant recipients. It was worth, however, much more than the personal outlay incurred by the individual members of this Association, to have settled once for all in an authoritative manner, such a perplexing and delusive anticipation.

As your agent, having fulfilled my obligations to the best of my ability, this report is respectfully submitted by

Yours Respectfully,

JAMES USHER,
New York City.